The Computer as Crucible

The Computer as Crucible

An Introduction to Experimental Mathematics

Jonathan Borwein
Keith Devlin

with illustrations by Karl H. Hofmann

CRC Press
Taylor & Francis Group
Boca Raton London New York

CRC Press is an imprint of the
Taylor & Francis Group, an **informa** business

AN A K PETERS BOOK

First published 2009 by A K Peters, Ltd.

Published 2019 by CRC Press
Taylor & Francis Group
6000 Broken Sound Parkway NW, Suite 300
Boca Raton, FL 33487-2742

© 2009 by Taylor & Francis Group, LLC
CRC Press is an imprint of the Taylor & Francis Group, an informa business

No claim to original U.S. Government works

ISBN-13: 978-1-56881-343-1 (hbk)

Library of Congress Cataloging-in-Publication Data

Borwein, Jonathan M.
     The computer as crucible : an introduction to experimental mathematics / Jonathan Borwein and Keith Devlin ; with illustrations by Karl H. Hofmann.
       p. cm.
     Includes bibliographical references and index.
     ISBN 978-1-56881-343-1 (alk. paper)
     1. Experimental mathematics. I. Devlin, Keith J. II. Title.

QA8.7.B67 2008
510–dc22

                                                                    2008022180

Visit the Taylor & Francis Web site at
http://www.taylorandfrancis.com

and the CRC Press Web site at
http://www.crcpress.com

For *Jakob Joseph*, age two,
and all others who will experience
much more powerful mathematical crucibles

Contents

Preface

Our aim in writing this book was to provide a short, readable account of experimental mathematics. (Chapter 1 begins with an explanation of what the term "experimental mathematics" means.) It is not intended as a textbook to accompany a course (though good instructors could surely use it that way). In particular, we do not aim for comprehensive coverage of the field; rather, we pick and choose topics and examples to give the reader a good sense of the current state of play in the rapidly growing new field of experimental mathematics. Also, there are no large exercise sets. We do end each chapter with a brief section called "Explorations," in which we give some follow-up examples and suggest one or two things the reader might like to try. There is no need to work on any of those explorations to proceed through the book, but we feel that trying one or two of them is likely to increase your feeling for the subject. Answers to those explorations can be found in the "Answers and Reflections" chapter near the end of the book.

This book was the idea of our good friend and publisher (plus mathematics PhD) Klaus Peters of A K Peters, Ltd. It grew out of a series of three books that one of us (Borwein) coauthored on experimental mathematics, all published by A K Peters: Jonathan Borwein and David Bailey's *Mathematics by Experiment* (2004); Jonathan Borwein, David Bailey, and Roland Girgensohn's *Experimentation in Mathematics* (2004); and David Bailey, Jonathan Borwein, Neil J. Calkin, Roland Girgensohn, D. Russell Luke, and Victor H. Moll's *Experimental Mathematics in Action* (2007).

We both found this an intriguing collaboration. Borwein, with a background in analysis and optimization, has been advocating and working in the new field of experimental mathematics for much of his career. This pursuit was considerably enhanced in 1993 when he was able to open the Centre for Experimental and Constructive Mathematics at Simon Fraser University, which he directed for a decade. (Many of the results presented here are due to Borwein, most often in collaboration with others, particularly Bailey.) Devlin, having focused on mathematical logic and set theory for the first half of his career, has spent much of the past twenty years looking at the emerging new field known as mathematical cognition, which tries to understand how the human brain does mathematics, how it acquires mathematical ability in the first place, and how mathematical thinking combines with other forms of reasoning, including machine computation. In working together on this book, written to explain to those not in the field what experimental mathematics is and how it is done, Borwein was on the inside looking out, and Devlin was on the outside looking in. We saw reassuringly similar scenes.

Experimental mathematics is fairly new. It is a way of doing mathematics that has been made possible by fast, powerful, and easy-to-use computers, by networks, and by databases.

The use of computers in mathematics *for its own sake* is a recent phenomenon—much more recent than the computer itself, in fact. (This surprises some outsiders, who assume, incorrectly, that mathematicians led the computer revolution. To be sure, mathematicians invented computers, but then they left it to others to develop them, with very few mathematicians actually using them until relatively recently.)

In fact, in the late 1980s, the American Mathematical Society, noting that mathematicians seemed to be lagging behind the other sciences in seeing the potential offered by computers, made a deliberate effort to make the mathematical community more aware of the possibilities presented by the new technology. In 1998, their flagship newsletter, the *Notices of the American Mathematical Society*, introduced a "Computers and Mathematics" section, edited originally by the late Jon Barwise, then (from October 1992 through December 1994) by Devlin. Devlin's interest in how the use of computers can change mathematical practice was part of his growing fascination with mathematical cognition. Correspondingly, Borwein's experience led to a growing interest in mathematical visualization and mathematical aesthetics.

A typical edition of the "Computers and Mathematics" section began with a commissioned feature article, followed by reviews of new mathematical software systems. Here is how Devlin opened his first "Computers and Mathematics" section: "Experimental mathematics is the theme of this month's feature article, written by the Canadian mathematical brothers Jonathan and Peter Borwein."

With this book, the circle is complete!

The "Computers and Mathematics" section was dropped in January 1995, when the use of computers in the mathematical community was thought to have developed sufficiently far that separate treatment in the

Notices was no longer necessary. As this short book should make abundantly clear, things have come a long way since then.

Both authors want to thank Klaus Peters for coming up with the idea for this book, and for his continued encouragement and patience over the unexpectedly long time it took us to mesh our sometimes insanely busy schedules sufficiently to make his vision a reality. And both authors are enormously grateful to Karl Heinrich Hofmann for generously providing his always entertaining and occasionally "subversive" illustrations, some of which reflect his Platonic ideas.

Jonathan Borwein

Keith Devlin

March 2008

What do I see here?

Chapter 1

⤳

What Is Experimental Mathematics?

I know it when I see it.
—Potter Stewart (1915–1985)

United States Supreme Court justice Potter Stewart famously observed in 1964 that, although he was unable to provide a precise definition of pornography, "I know it when I see it." We would say the same is true for experimental mathematics. Nevertheless, we realize that we owe our readers at least an approximate initial definition (of experimental mathematics, that is; you're on your own for pornography) to get started with, and here it is.

Experimental mathematics is the use of a computer to run computations—sometimes no more than trial-and-error tests—to look for patterns, to identify particular numbers and sequences, to gather evidence in support of specific mathematical assertions that may themselves arise by computational means, including search. Like contemporary chemists—and before them the alchemists of old—who mix various substances together in a crucible and heat them to a high temperature to see what happens, today's experimental mathematician puts a hopefully potent mix of numbers, formulas, and algorithms into a computer in the hope that something of interest emerges.

Had the ancient Greeks (and the other early civilizations who started the mathematics bandwagon) had access to computers, it is likely that the word "experimental" in the phrase "experimental mathematics" would be superfluous; the kinds of activities or processes that make a particular mathematical activity "experimental" would be viewed simply as *mathematics*. We say this with some confidence because if you remove from our initial definition the requirement that a computer be used, what would be

left accurately describes what most, if not all, professional mathematicians spend much of their time doing, and always have done!

Many readers, who studied mathematics at high school or university but did not go on to be professional mathematicians, will find that last remark surprising. For that is not the (carefully crafted) image of mathematics they were presented with. But take a look at the private notebooks of practically any of the mathematical greats and you will find page after page of trial-and-error experimentation (symbolic or numeric), exploratory calculations, guesses formulated, hypotheses examined (in mathematics, a "hypothesis" is a guess that doesn't immediately fall flat on its face), etc.

The reason this view of mathematics is not common is that you have to look at the private, *unpublished* (during their career) work of the greats in order to find this stuff (by the bucketful). What you will discover in their *published* work are precise statements of true facts, established by logical proofs that are based upon axioms (which may be, but more often are not, stated in the work).

Because mathematics is almost universally regarded, and commonly portrayed, as the search for pure, eternal (mathematical) truth, it is easy to understand how the published work of the greats could come to be regarded as constitutive of what mathematics actually *is*. But to make such an identification is to overlook that key phrase "the search for." Mathematics is not, and never has been, merely the end product of the search; the process of discovery is, and always has been, an integral part of the subject. As the great German mathematician Carl Friedrich Gauss wrote to his colleague Janos Bolyai in 1808, "It is not knowledge, but the act of learning, not possession but the act of getting there, which grants the greatest enjoyment."[1]

In fact, Gauss was very clearly an "experimental mathematician" of the first order. For example, in 1849 he recounted his analysis of the density of prime numbers [Goldstein 73]:

> I pondered this problem as a boy, in 1792 or 1793, and found that the density of primes around t is $1/\log t$, so that the number of primes up to a given bound x is approximately
>
> $$\int_2^x dt/\log t.$$

Formal proof that Gauss's approximation is asymptotically correct, which is now known as the Prime Number Theorem, did not come until 1896, more than 100 years after the young genius made his experimental discovery.

To give just one further example of Gauss's "experimental" work, we learn from his diary that, one day in 1799, while examining tables of integrals provided originally by James Stirling, he noticed that the reciprocal of the integral

$$\frac{2}{\pi} \int_0^1 \frac{dt}{\sqrt{1 - t^4}}$$

agreed numerically with the limit of the rapidly convergent arithmetic-geometric mean iteration (AGM):

$$a_0 = 1, \quad b_0 = \sqrt{2};$$

$$a_{n+1} = \frac{a_n + b_n}{2}, \quad b_{n+1} = \sqrt{a_n b_n}.$$

---

[1]The complete quote is: "It is not knowledge, but the act of learning, not possession but the act of getting there, which grants the greatest enjoyment. When I have clarified and exhausted a subject, then I turn away from it, in order to go into darkness again; the never-satisfied man is so strange if he has completed a structure, then it is not in order to dwell in it peacefully, but in order to begin another. I imagine the world conqueror must feel thus, who, after one kingdom is scarcely conquered, stretches out his arms for others."

The sequences (a_n) and (b_n) have the common limit

$$1.1981402347355922074\ldots.$$

Based on this purely computational observation (which he made to 11 places), Gauss conjectured and subsequently proved that the integral is indeed equal to the common limit of the two sequences. It was a remarkable result, of which he wrote in his diary, "[the result] will surely open up a whole new field of analysis." He was right. It led to the entire vista of nineteenth-century elliptic and modular function theory.

For most of the history of mathematics, the confusion of the activity of mathematics with its final product was understandable: after all, both activities were done by the same individual, using what to an outside observer were essentially the same activities—staring at a sheet of paper, thinking hard, and scribbling on that paper.[2] But as soon as mathematicians started using computers to carry out the exploratory work, the distinction became obvious, especially when the mathematician simply hit the ENTER key to initiate the experimental work, and then went out to eat while the computer did its thing. In some cases, the output that awaited the mathematician on his or her return was a new "result" that no one had hitherto suspected and might have no inkling how to prove.

The scare quotes around the word "result" in that last paragraph are to acknowledge that the adoption of experimental methods does not necessarily change the notion of mathematical truth, nor the basic premise that the only way a mathematical statement can be certified as correct is when a formal proof has been found. Whenever a relationship has been obtained using an experimental approach—and in this book we will give many specific examples—finding a formal proof remains an important and legitimate goal, although not the only goal.

What makes experimental mathematics different (as an enterprise) from the classical conception and practice of mathematics is that the experimental process is regarded not as a precursor to a proof, to be relegated to private notebooks and perhaps studied for historical purposes only after a proof has been obtained. Rather, experimentation is viewed as a signifi-

---

[2]The confusion would have been harmless but for one significant negative consequence: it scared off many a young potential mathematician, who, on being unable instantaneously to come up with the solution to a problem or the proof of an assertion, would erroneously conclude that they simply did not have a mathematical brain.

cant part of mathematics in its own right, to be published, to be considered by others, and (of particular importance) *to contribute to our overall mathematical knowledge.* In particular, this gives an epistemological status to assertions that, while supported by a considerable body of experimental results, have not yet been formally proved, and in some cases may never be proved. (As we shall see, it may also happen that an experimental process itself yields a formal proof. For example, if a computation determines that a certain parameter p, known to be an integer, lies between 2.5 and 3.784, that amounts to a rigorous proof that $p = 3$.)

When experimental methods (using computers) began to creep into mathematical practice in the 1970s, some mathematicians cried foul, saying that such processes should not be viewed as genuine mathematics—that the one true goal should be formal proof. Oddly enough, such a reaction would not have occurred a century or more earlier, when the likes of Fermat, Gauss, Euler, and Riemann spent many hours of their lives carrying out (*mental*) calculations in order to ascertain "possible truths" (many but not all of which they subsequently went on to prove). The ascendancy of the notion of proof as the sole goal of mathematics came about in the late nineteenth and early twentieth centuries, when attempts to understand the infinitesimal calculus led to a realization that the intuitive concepts of such basic concepts as function, continuity, and differentiability were highly problematic, in some cases leading to seeming contradictions. Faced with the uncomfortable reality that their intuitions could be inadequate or just plain misleading, mathematicians began to insist that value judgments were hitherto to be banished to off-duty chat in the mathematics common room and nothing would be accepted as legitimate until it had been formally proved.

This view of mathematics was the dominant one when both your present authors were in the process of entering the profession. The only way open to us to secure a university position and advance in the profession was to *prove* theorems. As the famous Hungarian mathematician Paul Erdős (1913–1996) is often quoted as saying, "a mathematician is a machine for turning coffee into theorems."[3]

---

[3] A more accurate rendition is: "Renyi would become one of Erdős's most important collaborators. ... Their long collaborative sessions were often fueled by endless cups of strong coffee. Caffeine is the drug of choice for most of the world's mathematicians and coffee is the preferred delivery system. Renyi, undoubtedly wired on espresso, summed this up in a

I see the use of machines to produce mathematics—even one that converts coffee into theorems!

As it happened, neither author fully bought into this view. Borwein adopted computational, experimental methods early in his career, using computers to help formulate conjectures and gather evidence in favor of them, while Devlin specialized in logic, in which the notion of proof is itself put under the microscope, and results are obtained (and published) to the effect that a certain statement, while true, is demonstrably not provable—a possibility that was first discovered by the Austrian logician Kurt Gödel in 1931.

What swung the pendulum back toward (openly) including experimental methods, we suggest, was in part pragmatic and part philosophical. (Note that word "including." The *inclusion* of experimental processes in no way eliminates proofs. For instance, no matter how many zeroes

---

famous remark almost always attributed to Erdős: '*A mathematician is a machine for turning coffee into theorems.*' ... Turan, after scornfully drinking a cup of American coffee, invented the corollary: '*Weak coffee is only fit for lemmas*'" [Schecter 98, p. 155].

of the Riemann zeta function are computed and found to have real part equal to 1/2, the mathematical community is not going to proclaim that the Riemann hypothesis—that all zeroes have this form—is true.[4])

The pragmatic factor behind the acknowledgment of experimental techniques was the growth in the sheer *power* of computers to search for patterns and to amass vast amounts of information in support of a hypothesis.

At the same time that the increasing availability of ever cheaper, faster, and more powerful computers proved irresistible for some mathematicians, there was a significant, though gradual, shift in the way mathematicians viewed their discipline. The Platonistic philosophy that abstract mathematical objects have a definite existence in some realm outside of humankind, with the task of the mathematician being to uncover or discover eternal, immutable truths about those objects, gave way to an acceptance that the subject is the product of humankind, the result of a particular kind of human thinking.

In passing, let us mention that the ancient-sounding term "Platonistic," for a long-standing and predominant philosophy of working mathematicians, is fairly recent. It was coined in the 1930s, a period in which Gödel's results made mathematical philosophers and logicians think very hard about the nature of mathematics. Mathematicians largely ignored the matter as of concern only to philosophers. In a similar vein, the linguist Steve Pinker recently wrote: "I don't think bio-chemists are going to be the least bit interested in what philosophers think about genes." This led biologist Steve Jones to retort: "As I've said in the past, philosophy is to science as pornography is to sex: It's cheaper, easier, and some people prefer it."[5]

It would be a mistake to view the Platonist and the product-of-the-human-mind views of mathematics as an exclusive either-or choice. A characteristic feature of the particular form of thinking we call mathematics is that it *can be thought of* in Platonistic terms—indeed most mathematicians report that such is how it appears and feels when they are actually doing mathematics.

---

[4]Opinions differ as to whether, or to what degree, the computational verification of billions of cases provides meaningful information as to how likely the hypothesis is to be true. We'll come back to this example shortly.

[5]This exchange can be found in *The Scientist*, June 20th, 2005.

The shift from Platonism to viewing mathematics as just another kind of human thinking brought the discipline much closer to the natural sciences, where the object is not to establish "truth" in some absolute sense, but to analyze, to formulate hypotheses, and to obtain evidence that either supports or negates a particular hypothesis.

In fact, as the Hungarian philosopher Imre Lakatos made clear in his 1976 book *Proofs and Refutations*, published two years after his death, the distinction between mathematics and natural science—as practiced—was always more apparent than real, resulting from the fashion among mathematicians to suppress the exploratory work that generally precedes formal proof.

By the mid-1990s, it was becoming common to "define" mathematics as a science—"the science of patterns" (an acceptance acknowledged and reinforced by Devlin's 1994 book *Mathematics: The Science of Patterns*).

The final nail in the coffin of what we shall call "hard-core Platonism" was driven in by the emergence of computer proofs, the first really major example being the 1976 proof of the famous Four Color Theorem, a statement that to this day is accepted as a theorem solely on the basis of an argument (actually, today at least two different such arguments) of which a significant portion is of necessity carried out by a computer.

The degree to which mathematics has come to resemble the natural sciences can be illustrated using the Riemann hypothesis, which we mentioned earlier. To date, the hypothesis has been verified computationally for the *ten trillion* zeroes closest to the origin. But every mathematician will agree that this does not amount to a conclusive proof. Now suppose that, next week, a mathematician posts on the Internet a five-hundred page argument that she or he claims is a proof of the hypothesis. The argument is very dense and contains several new and very deep ideas. Several years go by, during which many mathematicians around the world pore over the proof in every detail, and although they discover (and continue to discover) errors, in each case they or someone else (including the original author) is able to find a correction. At what point does the mathematical community as a whole declare that the hypothesis has indeed been proved? And even then, which do you find more convincing, the fact that there is an argument for which none of the hundred or so errors found so far have proved to be fatal, or the fact that the hypothesis has been verified computationally (and, we shall assume, *with total certainty*) for 10

Plato: Look, Ari, up there are arbitrarily long arithmetic sequences in the set of primes – and will be there when the sun has engulfed the earth!

Aristoteles: Listen, old man, the computer has driven the final nail into the coffin of your "ideas." Down here is where the action is as long as we and IT are here.

trillion cases? Different mathematicians will give differing answers to this question, but their responses are mere *opinions*.

In one fairly recent case, the editors of the *Annals of Mathematics* decided to publish a proof of a certain result with the disclaimer that after a committee of experts had examined the proof in great details for four years, the most positive conclusion they had been able to arrive at was that they were "99% certain" the argument was correct, but could not be absolutely sure. After other leading mathematicians intervened, the journal editors relented, and the paper was published without the disclaimer, but the point had been established: the mathematical world had changed.

The problematic proof was Thomas Hales's solution of the Kepler sphere packing problem [Hales 05]. It actually involved some computational reasoning, but the principle was established: given sufficient complexity, no human being can ever be certain an argument is correct, nor even a group of world experts. Hales's method ultimately relied on using a linear programming package that certainly gives correct answers but was never intended to certify them.

With a substantial number of mathematicians these days accepting the use of computational and experimental methods, mathematics has indeed grown to resemble much more the natural sciences. Some would argue that it simply *is* a natural science. If so, it does however remain, and we believe ardently will always remain, the most secure and precise of the sciences. The physicist or the chemist must rely ultimately on observation, measurement, and experiment to determine what is to be accepted as "true," and there is always the possibility of a more accurate (or different) observation, a more precise (or different) measurement, or a new experiment (that modifies or overturns the previously accepted "truths"). The mathematician, however, has that bedrock notion of proof as the final arbitrator. Yes, that method is not (in practice) perfect, particularly when long and complicated proofs are involved, but it provides a degree of certainty that no natural science can come close to. (Actually, we should perhaps take a small step backward here. If by "come close to" you mean an agreement between theory and observation to ten or more decimal places of accuracy, then modern physics has indeed achieved such certainty on some occasions.)

So what kinds of things does an experimental mathematician do? More precisely, and we hope that by now our reader appreciates the reason for this caveat, what kinds of activity does a *mathematician* do that classify, or can be classified, as "experimental mathematics"? Here are some that we will describe in the pages that follow:

1. symbolic computation using a computer algebra system such as Mathematica or Maple,

2. data visualization methods,

3. integer-relation methods, such as the PSLQ algorithm (see later),

4. high-precision integer and floating-point arithmetic,

5. high-precision numerical evaluation of integrals and summation of infinite series,

6. use of the Wilf-Zeilberger algorithm for proving summation identities (we're not doing that one),

7. iterative approximations to continuous functions (ditto),

8. identification of functions based on graph characteristics.

We should point out that our brief account in no way sets out to provide a comprehensive coverage of contemporary experimental mathematics. Rather, we focus on a particular slice through the field, by way of providing illustration of this powerful, and growing, new *approach* to mathematical discovery that the computer has made possible. (Though we should repeat our earlier observation that in days past, many of the greatest mathematicians spent many hours in "experimental pursuits," doing masses of computations with no aid other than a pen and paper and the power of their own intellect—or sometimes that of an assistant or two.)[6]

For the most part, our slice comprises (bits of) experimental real analysis and experimental analytic number theory, some of the former coming from problems in modern physics. In the final chapter (Chapter 11), we will provide a very brief survey of the use of experimental methods in some other parts of mathematics.

Explorations

One of the tantalizing things about computer experimentation is to learn how to distinguish when you might learn something by experimenting and when "messing about" is a waste of time. Ideally, you should run every experiment like a rigorous biology experiment with a null hypothesis,

---

[6]Until the second half of the twentieth century, the English word "computer" was used to refer to a human being, not a machine.

an experimental design, predetermined statistical tests, impeccable log-books (paper or digital), and so on. In reality there will always be messing about. So why not start with some.

1. *Recognizing sequences.* What rules generate the following?

    (a) 6, 28, 496, 8128, 33550336, 8589869056, 137438691328, ...

    (b) 1, 1, 2, 5, 15, 52, 203, 877, 4140, 21147, 115975, 678570, 4213597, ...

    (c) 1, 1, 2, 4, 9, 21, 51, 127, 323, 835, 2188, 5798, 15511, 41835, ...

    (d) 1, 2, 6, 22, 94, 454, 2430, 14214, 89918, 610182, 4412798, ...

    (e) 1, 4, 11, 16, 24, 29, 33, 35, 39, 45, 47, 51, 56, 58, 62, 64, ...

    (f) 1, 20, 400, 8902, 197281, 4865617, ...

2. *The 3n + 1 problem.* Some open problems may initially seem ripe for experimentation and their true level of difficulty only emerges later. A famous example is the "$3n + 1$" problem, which has many other names: Collatz's problem, the Syracuse problem, Kakutani's problem, Hasse's algorithm, and Ulam's problem. You start with the following simple algorithm that you apply recursively, starting with an arbitrary natural number: if n is even, divide it by 2; if n is odd, multiply it by 3 and add 1; continue until you come back to 1. The problem is: does this process always terminate?

   For example, if you start with 13, you get

   $$13 \rightarrow 40 \rightarrow 20 \rightarrow 10 \rightarrow 5 \rightarrow 16 \rightarrow 8 \rightarrow 4 \rightarrow 2 \rightarrow 1.$$

   Notice that the rule cycles indefinitely through the final subsequence here: 4, 2, 1. It is easy to write a simple program, and when you do, you will find that you always end up with 1. For some starting values it takes a large number of steps, and on the way you will encounter numbers that are very large, before finally starting to drop back to 1. Such sequences are sometimes called hailstone or juggler sequences.

   See what happens when you start with 7. (This one can be done quickly without recourse to a computer.) Then try some other starting values, say 27, which takes 111 steps.

What happens if you change the 3 to a 5 to give the "$5n + 1$" rule? You can also ask yourself how the $3n + 1$ conjecture could fail. To do so, there must be a starting value for which the sequence either diverges or settles into an infinite loop. Many variants do loop [Franco and Pomerance 95].

When you hear that the mathematician John Conway has shown that some problems like this are undecidable, and that the first 5×10^{13} cases of the $3n + 1$ rule are known to stop, you begin to get a sense of how complex the behavior of such seemingly simple rules can be.

3. *Continued fractions.* For irrational numbers, continued fractions provide an excellent source of many hours of exploratory work using a computer algebra system. But before you begin, it's wise to familiarize yourself with the abbreviated notation whereby $[a_0, a_1, a_2, a_3, \ldots, a_n, \ldots]$ abbreviates the more space-hungry expression

$$a_0 + \cfrac{1}{a_1 + \cfrac{1}{a_2 + \cfrac{1}{a_3 + \cdots + \cfrac{1}{a_n + \cdots}}}},$$

where $a_0, a_1, a_2, a_3, \ldots, a_n, \ldots$ are natural numbers.

If $\alpha = [a_0, a_1, a_2, a_3, \ldots, a_n, \ldots]$, the *partial quotients* a_k relate to the number α as follows. The continued fraction encodes the information that with initial conditions $q_0 := 1 =: p_{-1}, q_{-1} := -1, p_0 := a_0$, and with

$$p_{n+1} := a_{n+1} p_n + p_{n-1},$$
$$q_{n+1} := a_{n+1} q_n + q_{n-1},$$

there are very good rational approximations, $[a_0, a_1, \ldots, a_n] = p_n/q_n$, which tend to α. These *convergents* are easy to compute from the above recursion.[7]

The process of determining the partial quotients a_n from α is equally efficient: Let $\alpha_0 := \alpha$ and repeatedly compute

$$a_n = \lfloor \alpha_n \rfloor, \alpha_{n+1} = 1/(\alpha_n - a_n).$$

---

[7]See also http://mathworld.wolfram.com/ContinuedFraction.html.

Thus, a_k is the integer part[8] of α_k and α_{k+1} is its fractional part. The following code implements this in Maple:

```
cf:=proc(alpha,n) local a, k, r; a:=alpha; r:=alpha;
   for k to n do a:=a,trunc(r); r:=1/frac(r); od;
  a; end:
```

Implement the above code (or an equivalent one in your favorite programming language) and use it to compute the first ten partial quotients for π and for e.

---

[8]The "half" brackets stand for the *floor* function: $\lfloor x \rfloor$ is the largest integer less than or equal to x. Its counterpart, the *ceiling* function $\lceil x \rceil$, gives the smallest integer greater than or equal to x. For example, $\lfloor \pi \rfloor = 3$ and $\lceil \pi \rceil = 4$.

$1 \, ly = 9.4607 \times 10^{15} \, m$

What is in a quadrillion?
A tenth of a light year
is a quadrillion meters.

Chapter 2

⤳

What Is the Quadrillionth Decimal Place of π?

We may never know the answer to the title question. Although there are some efficient algorithms for computing π, they all require computing the decimal digits sequentially, 3.14159 etc., and even with the most powerful computers available, it would simply take too long to reach the quadrillionth place. A quadrillion is 10^{15}. At the time of writing, π has been computed to just over 1.25 trillion decimal places. For the record, the ten decimal places ending in position one trillion are 6680122702. We'll come back later (in Chapter 7) to efforts to determine ever more of the decimal expansion of π and describe some of the methods used, but for now we want to look at a slightly different question: what is the quadrillionth place in the *binary* expansion of π?

Your initial guess is probably that this is equally hopeless, but this turns out not to be the case. In fact, we'll tell you the answer: the quadrillionth binary digit in π is 0.[1] While we are on it, we'll let you in on another secret: you could be the first person to compute the quadrillionth place in the *hexadecimal* expansion of π; the method we shall describe works for both number-bases, 2 and 16, but no one has yet used it to find the quadrillionth hexadecimal place. In terms of its expansion, π would be a lot easier to handle if humans had evolved to have two fingers or sixteen!

The key to this remarkable (albeit seemingly totally useless) piece of knowledge is the following formula, discovered in 1995 by Peter Borwein

---

[1]See the end of the chapter for details about this particular calculation.

(the brother of your first author), David Bailey, and Simon Plouffe, and named the BBP formula after them:

$$\pi = \sum_{k=0}^{\infty} \frac{1}{16^k} \left(\frac{4}{8k+1} - \frac{2}{8k+4} - \frac{1}{8k+5} - \frac{1}{8k+6} \right) \tag{2.1}$$

[Bailey et al. 97]. (It is the nature of much experimental mathematics to date that it involves formulas that occupy a fair amount of space on the page. The above formula is but a gentle introduction to what is to follow. In most of the examples we present, however, the formulas involve only very simple notions, as is the case here.)

Using formula (2.1), you can directly calculate binary or hexadecimal digits of π beginning with the nth digit, without having to first compute the previous $n - 1$ digits. All you need to carry out the computation is a simple algorithm using standard 64-bit or 128-bit arithmetic. We'll describe how this calculation is carried out, but our real interest is in how the BBP formula came to be discovered.

The story began with the well-known classical formula

$$\log 2 = \sum_{k=1}^{\infty} \frac{1}{k2^k}. \tag{2.2}$$

Around 1994, Peter Borwein and Simon Plouffe of Simon Fraser University in Canada realized that you could use this formula to calculate individual binary digits of $\log 2$. Suppose you want to compute a few binary digits starting at position $d + 1$ for some positive integer d. This is equivalent to computing $\{2^d \log 2\}$, where $\{\ldots\}$ denotes fractional part. From (2.2) we get

$$\{2^d \log 2\} = \left\{ \left\{ \sum_{k=0}^{d} \frac{2^{d-k}}{k} \right\} + \left\{ \sum_{k=d+1}^{\infty} \frac{2^{d-k}}{k} \right\} \right\}$$

$$= \left\{ \left\{ \sum_{k=0}^{d} \frac{2^{d-k} \bmod k}{k} \right\} + \left\{ \sum_{k=d+1}^{\infty} \frac{2^{d-k}}{k} \right\} \right\}. \tag{2.3}$$

(Take this one step at a time; there is nothing deep going on here, just notational complexity. In calculating the fractional part of a sum, we can discard whole number parts at any stage we want, and we can insert "mod k" in the numerator of the first term because we are only interested in the fractional part of the quotient on division by k.)

Now let's see how we can use formula (2.3) to compute binary digits of log 2 starting at place $d + 1$. First, there is a highly efficient way to compute the numerators in the first sum, namely 2^{d-k} mod k. The naïve way would be to multiply 2 by itself $d - k$ times, discarding whole multiples of k whenever they arise. This would not require the ongoing storage of any number greater than k, but if $d - k$ is very large (as would occur if d were a quadrillion) it would require a great many steps—too many in fact. But there is a huge reduction in computation if you build up the power 2^{d-k} by iterated squaring. For example,

$$2^{50} = ((((2^2)^2)^2)^2)^2 \times (((2^2)^2)^2)^2 \times 2^2$$

requires just five doublings (plus a couple of multiplications) rather than fifty. If you were doing this modulo some fairly small modulus, say mod 10, then by discarding multiples of 10 as soon as they arise you could even carry out the calculation in your head. Some fairly straightforward and simple coding will easily produce a computer program that carries out such computations with great efficiency.[2]

This provides an efficient way to compute the finite first sum in (2.3). Since the aim is to compute just a few binary places starting at place $d + 1$, the second (infinite) sum can be truncated after just a few terms. (Notice that the individual terms in the second sum rapidly become small.) And there is your answer.

If $d < 10^7$, the entire computation can be carried out using the 64-bit arithmetic that comes standard in many computers. With a 128-bit floating-point arithmetic unit, you can comfortably handle $d \leq 10^{15}$. For d beyond 10^{15}, you would need special arithmetic routines, but that takes you beyond the quadrillionth place that was our starting point!

Well, that was a cute observation, but so far nothing remotely experimental. But as soon as they had made their log 2 discovery, Borwein and Plouffe asked themselves whether it was possible to pull off a similar stunt

---

[2]This parenthetic gymnastics is typical of modern computation. There is often a smart reorganization of a computation resulting in a huge saving of work.

for π. That might be no less useless to humankind, but given π's status in mathematics, and the fascination with computing the expansion of π going back to the ancient Greeks, it sure would be interesting![3]

It is obvious that the same technique can be applied to any constant α that can be written using a formula of the form

$$\alpha = \sum_{n=1}^{\infty} \frac{p(n)}{b^n q(n)},$$

where $b > 1$ is an integer and p and q are polynomials with integer co-efficients, and where q has no zeroes at positive integer arguments. Although there are several series expansions of π, a quick search of the literature revealed nothing of this form to Borwein and Plouffe. So the two joined forces with David Bailey, a computational mathematician based at the time at NASA Ames Research Center in California,[4] to see if π could be expressed as a linear combination of other constants of this form.

Bailey had developed a computer program that could find such linear combinations, based on *PSLQ algorithm*, developed by the American mathematician and sculptor Helaman Ferguson. The name "PSLQ" comes from the approach the algorithm uses, which involves a partial-sum-of-squares vector and a LQ (lower-diagonal-orthogonal) matrix.

The PSLQ algorithm is an example of what is known as an integer relation algorithm. Here, in general terms, is how such algorithms work.

---

[3]Of course, declaring a particular mathematical result "useless" depends on what exactly you mean by "useful," and even then is a value judgment that history may prove to be wrong. Giving pleasure to a great many people or stimulating them to think about the result could surely be classified as "useful," and that would make the Borwein-Plouffe result "useful" in the same way that literature and art are "useful." In the case of mathematics, there is often a hidden utility in that the methods developed to obtain the "useless" result turn out to have other applications in decidedly real-world settings. And in fact, the algorithm has proved useful in the sense that it has been built into at least one Fortran compiler because of its low storage requirements. It has also been used to confirm the record computation of π to one billion hexadecimal places by computing a string of digits around the trillionth place. This took hours instead of the months required to recompute all the digits on a large parallel system. Details may be found in the book *Mathematics by Experiment* [Borwein and Bailey 08]. Like many mathematicians, your two authors do not study mathematics because of its "utility," but we recognize that the question of utility is of interest to many.

[4]He is now at Lawrence-Berkeley Laboratory in California.

First, note that expressing a given real constant a as a rational linear combination,

$$a = q_1 a_1 + q_2 a_2 + \ldots + q_n a_n,$$

of certain other real constants $a_1, a_2, \ldots, a_n$ (where the coefficients $q_1, \ldots, q_n$ are rational numbers) is equivalent to finding integer coefficients $\lambda_0, \lambda_1, \ldots, \lambda_n$, such that $\lambda_0 \neq 0$ and

$$\lambda_0 a + \lambda_1 a_1 + \ldots + \lambda_n a_n = 0.$$

Given any real constants $a_0, a_1, \ldots, a_n$, and a preassigned degree of precision ε, an integer relation algorithm finds integer coefficients $\lambda_0, \lambda_1, \ldots, \lambda_n$ such that

$$|\lambda_0 a_0 + \lambda_1 a_1 + \ldots + \lambda_n a_n| < \varepsilon,$$

or else it tells you that no such expression exists within a ball of some given radius about the origin.

Note that you are rarely going to get 0 as a computed answer for the sum, and even if you do, you cannot be sure that this is not just an artifact of the inherent precision of the computer's arithmetic unit, which computes only to a pre-set number of binary places. But, by setting the precision parameter ε sufficiently small, you can achieve as much confidence as you require. For the purposes of experimental mathematics, this is generally sufficient to establish "experimentally equal to 0."

Does this actually prove that the linear combination is 0 (or that there *is* a linear combination that equals 0)? Of course not. Yet, many (most?) mathematicians, if presented with two closed-form expressions that produce the same decimal expansion when computed to, for example, 100 (or 500, or 1000, etc.) places, would conclude with considerable confidence that the two expressions were equal. That confidence would almost certainly be strong enough that they would be motivated to expend considerable time and effort trying to find a rigorous proof of equality.[5]

Minus all the details, here is how the PSLQ algorithm works.

Let x be the real input vector $(a_1, \ldots, a_n)$. The idea is to construct a series of matrices A_k such that the entries of the vector $y_k = A_k^{-1} x$ steadily decrease in size. This is done in such a way that, for any given iterate, the largest and smallest entries of y_k usually differ by no more than two or three orders of magnitude.

---

[5] But see Chapter 10 for some cases where this intuition goes wrong.

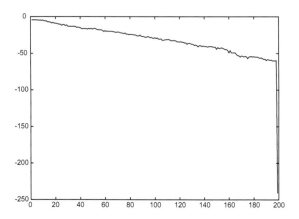

Figure 1. Plot of $\log_{10}(\min_i |x_i|)$ in a typical PSLQ run as a function of the iteration number, where x is the input vector.

When the algorithm detects a relation, the smallest entry of the y_k vector abruptly decreases to the order of the arithmetic precision being used (i.e., to 10^{-p}, where p is the precision level in digits). The desired integer relation is then given by the corresponding column of A_k^{-1}. Figure 1 shows this behavior for a typical PSLQ computation.

To be *confident* that the result the algorithm returns is a genuine integer relation (total certainty is not possible, of course), one might set the detection threshold ε to be, say, 10^{-100}. The precision level for the arithmetic used in the computation must then be set a few orders of magnitude smaller, to ensure that the unavoidable approximations in the computation do not significantly affect the result. (You almost always have to use high-precision arithmetic to run an integer relation algorithm.)

To return to Peter Borwein and Simon Plouffe's search for a formula for computing arbitrary binary digits of π, recall that they knew their $\log 2$ method would work for any constant of the form

$$\alpha = \sum_{k=0}^{\infty} \frac{p(k)}{q(k)2^k},$$

where p and q are integer polynomials with $\deg p < \deg q$ and q having no zeroes at nonnegative integer arguments. An initial search through the literature turned up around 25 constants of this type. But π was not one of them. However, they would be able to compute arbitrary binary digits for any number that can be expressed as a rational linear combination of such constants. Could they find such a linear combination for π?

When we left the two researchers, they had just turned to David Bailey and his implementation of the PSLQ integer relation algorithm in high-precision, floating-point arithmetic. Could Bailey's program find an integer relation for the real vector $(\alpha_1, \alpha_2, \ldots, \alpha_n)$, where $\alpha_1 = \pi$ and $\alpha_2, \ldots, \alpha_n$ are the known constants of the requisite form gleaned from the literature, each computed to several hundred decimal digit precision.

At first it couldn't. But what if they could find some more constants of the requisite type?

And so the search began. It wasn't exactly blind search, but it was close. And it went on for quite a while, with numerous restarts each time an additional formula was found in the literature.

But then, after a couple of months of computation, Bailey's program finally hit the mother lode. The actual formula it found was

$$\pi = {}_2F_1\left(\begin{array}{c} \frac{1}{4}, \frac{5}{4} \\ 1 \end{array} \Big| \frac{-1}{4}\right) + 2\arctan\left(\frac{1}{2}\right) - \log 5,$$

where the first term on the right is a hypergeometric function evaluation with value $0.955933837\ldots$. (See equation (4.2) to come.) Reducing this expression to sums gave the now famous BBP formula. The search was over.

We should point out that even with the BBP formula, computing the quadrillionth binary digit of π, the result of which we noted at the beginning of the chapter, is still a formidable calculation. It was carried out in September 2000, and was organized by Colin Percival, an undergraduate student at Simon Fraser University in Canada. It took 250 CPU-years, spread across 1,734 machines in 56 countries. Computation of the quadrillionth hexadecimal digit would be considerably harder, since it corresponds to the binary digit in place 4×10^{15}, though increases in computing power since Percival's computation have almost certainly made this feasible.

Finally, we'll end with an intriguing observation of Bailey and Richard Crandall. Consider the iteration:

$$x_0 = 0;$$
$$x_n = \left\{ 16x_{n-1} + \frac{120n^2 - 89n + 16}{512n^4 - 1024n^3 + 712n^2 - 206n + 21} \right\}.$$

The fraction in this formula comes from the BBP formula, (2.1). You simply combine the four fractions in (2.1) into one and shift the index by one.
Define y_n by

$$y_n = \lfloor 16x_n \rfloor.$$

Bailey and Crandall were led to look at the numbers y_n when they were analyzing the statistical distribution of the hexadecimal digits of π. If you divide the unit interval into 16 equal subintervals labeled $0, 1, \ldots, 15$, then y_n is the label of the subinterval in which x_n lies.

Now comes the surprising part. Computation of the sequence (y_n) has shown that the first million values are *exactly* the first million hexadecimal digits of $\pi - 3$. (This is a fairly difficult computation, requiring roughly n^2 bit-operations, and is not particularly amenable to parallel computation.) This led Bailey and Crandall to conjecture that the sequence (y_n) precisely generates the hexadecimal expansion of $\pi - 3$.

So far, this conjecture has not been definitively proved. But never mind proved; is it true? Or is the identity of the first million values a misleading accident, an example of the ever-present danger in experimental mathematics—a danger that is real for all that it is relatively rare—that patterns sometimes persist for hundreds, millions, billions, and occasionally *much* further, before breaking down? As we indicated earlier, we'll look at this worrisome, but intriguing, phenomenon of misleading evidence in Chapter 10.

Explorations

1. *BBP formulas.* These exist for a large variety of numbers in various bases. For example, you can find BBP formulas for:

   (a) the log of every prime less than 23,
   (b) π^2 in base 2 and base 3,

(c) Catalan's constant

$$G = \sum_{n=0}^{\infty} \frac{(-1)^n}{(2n+1)^2}$$

in base 2.

It has been shown conclusively that there is no similar formula for π in base 10.

2. *There is no known BBP formula for e in any base.* It has been conjectured that there is none. Can you find a "natural," slow series for e? (A series does not have to be rapidly convergent to be useful. As we have seen, it is the existence of an appropriate slow series that allows you to pick off digits.) As with Judge Potter, you will recognize "natural" when you see it.

3. *Spigot algorithms for π and e.* A *spigot method* for a numerical constant is one that can produce digits one by one ("drop by drop").

   This is especially easy for e, since carries are not a big issue. The following algorithm, due to S. Rabinowitz and S. Wagon, generates digits of e:

   Initialize an array A of length $n + 1$ to 1. Then repeat the following $(n - 1)$ times:

   (a) Multiply each entry in A by 10.
   (b) Starting from the right, reduce the ith entry of A modulo $i + 1$, carrying the quotient of the division one place left. The final quotient produced is the next digit of e.

   This algorithm is based on the following formula, which is a simple restatement of the fast series $e = \sum_{n=0}^{\infty} 1/n!$:

   $$e = 1 + \frac{1}{1}\left(1 + \frac{1}{2}\left(1 + \frac{1}{3}\left(1 + \frac{1}{4}\left(1 + \frac{1}{5}(1 + \cdots)\right)\right)\right)\right).$$

   Now implement a corresponding spigot algorithm for π, based on showing that

   $$\pi = 2 + \frac{1}{3}\left(2 + \frac{2}{5}\left(2 + \frac{3}{7}\left(2 + \cdots\left(2 + \frac{k}{2k+1} + \cdots\right)\right)\right)\right).$$

The last term can be approximated by $2 + 4k/(2k + 1)$, where $k = n \log_2 10$, to produce n digits of π "drop by drop."

If you want to run the algorithm without a specified end, you must take more care.

We can view the conjectural hex iteration for π that we finished the chapter with as a spectacular, albeit unproven, spigot algorithm for π in base 16. Indeed, if you heuristically assume that the hexadecimal tails of π can be treated as independent, uniformly distributed random variables in $[0, 1]$, then the probability of any further error can be shown to be less that one part in 10^8.

Chapter 3

⤳

What Is That Number?

Who was that masked man?
—The Lone Ranger

What is the following number?

$$3.1415926535897932385$$

The only *correct*, simple answer you can give is that it is the number 3.1415926535897932385. But most likely you read the question as asking for something more, something along the lines of "Can you provide a closed form that, to 20 places, expands to this decimal?" And almost certainly, you gave the answer π.

Here's another easy one: can you give a closed-form expression that, to 20 places, yields the following decimal?

$$0.7182818284590452354$$

If you recall the notation for fractional part we introduced in the last chapter, you can answer this question with $\{e\}$. Or, you could give the answer as $e - 2$.

Now try one more. Can you find a closed form that gives the following to 20 decimal places?

$$4.5007335678193158562$$

This one is not so easy. From the context, you might be able to find it after a bit of trial and error, but you probably have better things to do with your time. The answer is

$$\pi + \frac{1}{2} e.$$

Often in experimental mathematics, you carry out a calculation and produce a number to a certain number of decimal places, and you want to find a closed-form expression that yields that number to that accuracy, or maybe you just want to know if there is such a closed-form expression. The more decimal places you have, the more likely you are to suspect, or indeed believe, that the number you have been working with is actually the number given by that closed-form expression. Often such justified faith is a great spur to the discovery of a proof. Sometimes, further experimentation actually guides the proof process.

Even for simple examples like the third one above ($\pi + \frac{1}{2} e$), trial and error is obviously not an efficient strategy. (We could have been mean and made the third example $\pi + 0.499999e$.) But such a task is ideal for a fast computer.

Do you want to know what

$$62643383279502884197$$

might be? Using a search engine, a computer could quickly search a database of known mathematical constants and within moments tell you that it is 20 places of the decimal expansion of π starting at the 20th place.

There are publicly available websites that provide such a resource for free. The most popular is the On-Line Encyclopedia of Integer Sequences, developed and maintained by Neil Sloane at AT&T (with the help of a small army of highly-qualified assistants), which can be found at http://www.research.att.com/~njas/sequences/index.html. Type in the above sequence of twenty digits (separated by commas) and the encyclopedia will at once return the beginning of the decimal expansion of π with the entered sequence highlighted, together with a list of references to find information about π.[1]

The integer sequences look-up site is updated regularly, and asks users who have a sequence that is not in the database to send it in for inclusion. At the time of writing (January 2008), the site's database contains 135,307

---

[1] http://www.lbl.gov/wonder/bailey.html contains more such information and links to the site http://pi.nersc.gov, which lets you search for patterns—such as your phone number in the first four billion digits of π!

sequences. It started twenty years ago with a book containing 5,000 sequences. Implemented as a computer program, it is much more powerful; in addition to very rapid search, it can, for example, automatically tell you if twice your sequence is a known sequence, or if your sequence is a subsequence of a known sequence.

When look-up tables of known integer sequences and decimal expansions are combined with integer relation algorithms like the one described in the previous chapter, you have some extremely powerful machinery to carry out (experimental) mathematical investigations.

A free, publicly available resource for carrying out such investigations is the Inverse Symbolic Calculator (ISC), developed in the mid-1990s and maintained initially at the Centre for Experimental and Constructive Mathematics in the Department of Mathematics at Simon Fraser University in Canada, http://oldweb.cecm.sfu.ca/projects/ISC/ISCmain.html, and more recently in an updated parallelized form, ISC+, at Dalhousie University, http://ddrive.cs.dal.ca/~isc/. This resource will, for example, inform you that 19.999099979 is probably $e^{\pi} - \pi$.

Similar functionality is provided in the commercial mathematical software products Maple (the `identify` command) and, in a more limited version, Mathematica (the `Recognize` command, although this routine only recognizes algebraic numbers, and will make no progress with 19.999099979). Indeed, the ISC+ relies on careful exploitation of `identify` and similar tools.

Here is an excellent example of the effective use of a look-up table. In 1988, a gentleman by the name of Joseph Roy North of Colorado Springs was examining the Gregory series for π,

$$\pi = 4 \sum_{k=1}^{\infty} \frac{(-1)^{k+1}}{2k - 1} = 4 \left(1 - \frac{1}{3} + \frac{1}{5} - \frac{1}{7} + \cdots \right).$$

He noticed that, when this series was truncated to 5,000,000 terms, it gives a value that differs strangely from the true value of π. Here is the truncated Gregory value and the true value of π with the differences indicated:

3.14159**2**45358979323846464338327950**278**41971693993**873**058209749**4182**230781640...

3.14159**2**65358979323846264338327950**288**41971693993**751**058209749**4459**230781640...

| 2 | −2 | 10 | −122 | 2770 |

We've actually marked the differences in the fashion that emerged when Borwein analyzed the issue after North brought the problem to his attention. Let's see what is going on.

The series value differs, as one might expect from a series truncated to 5,000,000 terms, in the seventh decimal place—there is a 4 where there should be a 6. But then the next 13 digits are correct! Then, following another erroneous digit, the sequence is once again correct for an additional 12 digits. In fact, of the first 46 digits, only four differ from the corresponding decimal digits of π. Moreover, the erroneous digits appear to occur in positions that have a period of 14. Surely, there has to be an explanation.

A good way to start an investigation is to see if something similar happens with another series expansion, for example, the logarithm

$$\log 2 = 1 - \frac{1}{2} + \frac{1}{3} - \frac{1}{4} + \dots$$

And indeed it does, as the following value obtained by truncating the series shows:

0.69314708055995530941723212125817656807551613436025525140068000949418722...

0.69314718055994530941723212145817656807550013436025525412068000949339362...

| 1 | −1 | 2 | −16 | 272 | −7936 |

Once again, the erroneous digits appear in locations with a period of 14. In the first case, the differences from the correct values are $(2, -2, 10, -122, 2770)$, while in the second case the differences are $(1, -1, 2, -16, 272, -7936)$. Note that each integer in the first set is even; dividing by 2, we obtain $(1, -1, 5, -61, 1385)$.

Now we turn to Sloane's Internet-based Encyclopedia of Integer Sequences. This tool has no difficulty recognizing the first sequence as "Euler numbers" and the second as "tangent numbers."[2] Euler numbers and tangent numbers are defined in terms of the Taylor series for $\sec x$ and $\tan x$, respectively:

---

[2] When Borwein and his brother did the work presented here, the Encyclopedia only existed as a printed book, and while it contained the Euler numbers it did not include the sequences times 2. Today the online version takes care of such details automatically.

$$\sec x = \sum_{k=0}^{\infty} \frac{(-1)^k E_{2k} x^{2k}}{(2k)!},$$

$$\tan x = \sum_{k=0}^{\infty} \frac{(-1)^{k+1} T_{2k+1} x^{2k+1}}{(2k+1)!}.$$

This provides the key clue to the resolution of the mystery. We note that the following asymptotic expansions hold:

$$\frac{\pi}{2} - 2 \sum_{k=1}^{N/2} \frac{(-1)^{k+1}}{2k-1} \approx \sum_{m=1}^{\infty} \frac{E_{2m}}{N^{2m+1}},$$

$$\log 2 - \sum_{k=1}^{N/2} \frac{(-1)^{k+1}}{k} \approx \frac{1}{N} + \sum_{m=1}^{\infty} \frac{T_{2m-1}}{N^{2m}}.$$

Now the genesis of the anomaly is clear: North, in computing π by Gregory's series, had by chance truncated the series at 5,000,000 terms, which is exactly one-half of a fairly large power of 10. Indeed, setting $N = 10,000,000$ in the first of the above two asymptotic expansions shows that the first hundred or so digits of the truncated series value are small perturbations of the correct decimal expansion for π. Similar phenomena occur for other constants.[3] Mystery solved.

Here is one final example of the use of a look-up table. Suppose you are faced with finding a closed form for the sequence that starts like this:

$$1, \ -\frac{1}{3}, \ \frac{1}{25}, \ -\frac{1}{147}, \ \frac{1}{1089}, \ -\frac{3}{20449}, \ \frac{1}{48841}, \ -\frac{1}{312987},$$

$$\frac{25}{55190041}, \ -\frac{1}{14322675}, \ \frac{1}{100100025}, \ -\frac{49}{32065374675}, \ \dots$$

What was that you said? "You can't imagine why you would ever want to do such a thing?" Well, in the following chapter, we'll explain just what you might have been doing to reach such a point—this example is not something we just made up, it has a mathematical origin! For the moment, however, let's see how you might set about it.

---

[3]If we were working in hexadecimal, we would examine one-half of a fairly large power of 16.

A good first step is to try factorizing those increasingly daunting-looking denominators and see if any kind of pattern emerges. (Knowing that a particular sequence comes from a mathematically-meaningful or a real-world problem usually leads us to expect there to be a pattern, and it's just a matter of finding it. This supposition is, of course, rife with philosophical, psychological, and sociological considerations.) Here is what you get when you factorize both the numerators and denominators of the first eight terms—something that, at least for relatively small numbers, is easily performed on a computer, using standard-issue routines:

$$1, \frac{-1}{3}, \frac{1}{(5)^2}, \frac{-1}{(3)(7)^2}, \frac{1}{(3)^2(11)^2}, \frac{(-3)}{(11)^2(13)^2}, \frac{1}{(13)^2(17)^2},$$

$$\frac{-1}{(3)(17)^2(19)^2}, \frac{(5)^2}{(17)^2(19)^2(23)^2}.$$

Given all those squared terms, a natural next step might be to separate out the even and odd terms (on account of the alternating signs) and take the square roots. For the square roots of the even terms, this yields

$$1, \frac{1}{(5)}, \frac{1}{(3)(11)}, \frac{1}{(13)(17)}, \frac{(5)}{(17)(19)(23)}, \frac{1}{(3)(5)(23)(29)},$$

$$\frac{1}{(3)(5)^2(29)(31)}, \frac{(3)}{(29)(31)(37)(41)},$$

and after factoring out -3, the square roots of the odd terms start out

$$1, \frac{1}{(7)}, \frac{(3)}{(11)(13)}, \frac{1}{(17)(19)}, \frac{1}{(5)(19)(23)}, \frac{(7)}{(5)(23)(29)(31)},$$

$$\frac{1}{(3)(29)(31)(37)}, \frac{(3)}{(31)(37)(41)(43)}.$$

It is now apparent that both sequences have structure modulo six. Indeed the largest value is of the form $6n \mp 1$, except in the even case of 35 and the odd case of 25 which are not prime. Were the modular pattern not so clear we could produce more cases.

Given the pattern of the ascending $(6n \mp 1)$ terms, the next thing we might try is to express the fractions in terms of factorials. Consider the even terms. Multiplying by $(6n)!$ leads rapidly to enormous integers, so that doesn't help. But when you multiply by the central binomial coefficient $\binom{6n}{3n}$, you get the sequence

$$1, 4, 28, 220, 1820, 15504, 134596, 1184040, \ldots$$

Entering this into Sloane's *Encyclopedia of Integer Sequences* returns a single answer:

$$\binom{4n}{n}.$$

Thus, the even terms of the original sequence appear to be

$$s_{2n} = \left(\binom{4n}{n} \Big/ \binom{6n}{3n}\right)^2.$$

A similar process yields the following closed form for the odd terms:

$$s_{2n+1} = -\frac{1}{3}\left(\binom{6n+1}{3n} \Big/ \binom{4n+1}{n}\right)^2.$$

So there you have it. Our argument made significant use of computer technology, but it was by no means merely mindless key pushing.

Explorations

1. *Name that number.* Here are eight numbers you are challenged to identify from 20 digits or so. All can be discovered from at least one of the ISC, Sloane's encyclopedia, identify or Recognize:

   (a) 3.1462643699419723423

   (b) 2.9919718574637504583

   (c) 24.140692632779269007

   (d) 20.459157718361045475

   *(e) 8.409338623762925685

   (f) 1.3247179572447460260

   *(g) 1.1762808182599175065

   *(h) 0.69777465796400798203

   The three marked with an asterisk are likely to be harder work.

2. *Name that sum.* Identify

$$\sum_{n=0}^{\infty} r(n)^7 (1 + 14n + 76n^2 + 168n^3) \left(\frac{1}{8}\right)^{2n},$$

where

$$r(n) := \frac{\frac{1}{2} \cdot \frac{3}{2} \cdots \frac{(2n-1)}{2}}{n!} = \frac{\Gamma\left(n + \frac{1}{2}\right)}{\sqrt{\pi}\,\Gamma(n+1)}.$$

Here,

$$\Gamma(x) := \int_0^\infty t^{(x-1)} e^{-t} dt$$

(the Gamma function) is the unique function on the positive real numbers satisfying the functional equation

$$x\Gamma(x) = \Gamma(x+1), \quad \Gamma(1) = 1,$$

whose logarithm is convex. In particular, this means $\Gamma(n+1) = n!$ and that Γ is the only reasonable function interpolating the factorial to non-integer values.

You can compute the numerical value of the above summation quickly using Maple or Mathematica, since the series converges quite rapidly.

Wayne at work

Chapter 4

~~→

The Most Important Function
in Mathematics

There is more chance of him proving Riemann's hypothesis than wearing a sarong.

—*Guardian* (UK)

We promised we would explain the origin of that peculiar sequence of fractions we investigated in the previous chapter. The story provides yet another interesting episode in experimental mathematics.

Doubtless, your evaluation of the mathematical significance of the previous example will rocket sky high when we tell you that it came from work on the Riemann zeta function, arguably the most important function in all of mathematics.[1]

First, let us recall some basic facts about the zeta function. In a famous 1859 memoir [Riemann 59], the German mathematician Georg Friedrich Bernhard Riemann introduced the zeta function, which may be defined on positive integers $n > 1$ by

$$\zeta(n) = \sum_{k=1}^{\infty} \frac{1}{k^n}.$$

The zeta function may be extended to an analytic function defined almost everywhere on the complex plane by a process called analytic continuation. In his memoir, Riemann put forward the hypothesis that all zeroes

---

Epigraph: Headline in the sports section of the UK's *Guardian* newspaper, June 24, 2004, referring to British soccer star Wayne Rooney. The article contrasted the down-to-earth Rooney, the UK's leading soccer player, with David Beckham, his glamorous predecessor who, in addition to scoring goals, often did double duty as a fashion model. This is almost certainly the first time the Riemann hypothesis was used in a sports headline.

[1]All right, we admit that statements such as this, while guaranteed attention grabbers, are absurd. But there is no denying that the zeta function is extremely important in many areas of mathematics.

of $\zeta(s)$ for complex numbers $s = \sigma + i\gamma$ with $0 \le \sigma \le 1$ have $\sigma = 1/2$. This is the famous *Riemann hypothesis*, a proof of which has eluded the best mathematicians (and more recently, apparently, Wayne Rooney) for nearly 150 years.

Riemann's 1859 memoir did not contain any clues as to how he was led to make this conjecture. For many years mathematicians believed that Riemann had come to this conclusion on the basis of some profound intuition. Indeed, the Riemann hypothesis was held up as a premier example of the heights one could attain by sheer intellect alone. In 1929, however, long after Riemann's death, the renowned number theorist Carl Ludwig Siegel (1896–1981) learned that Riemann's widow had donated his working papers to the Göttingen University library. Among these papers, Siegel found several pages of dense numerical calculations, with a number of the lowest-order zeroes of the zeta function calculated to several decimal places each. One can only imagine that, were computers available in Riemann's time, the great German mathematician would have calculated several hundred zeroes. As it was, it is remarkable that he was able to formulate his conjecture on the basis of relatively little numerical evidence, but it seems clear that his method was one of experimental mathematics!

Our interest here is in the zeta function defined on integer arguments $s > 1$. One of the first obvious questions to ask is, what are its values? The answer is different according to whether the argument is even or odd, with the even arguments being by far the easier of the two cases to answer.

For any positive integer n,

$$\zeta(2n) = C_n \pi^{2n},$$

where C_n is a rational number. The first few values of the constants C_n are $C_1 = 1/6$, $C_2 = 1/90$, $C_3 = 1/945$, $C_4 = 1/9450$, $C_5 = 1/93555$.

In 1739, Leonhard Euler found a general expression for all C_n in terms of the Bernoulli numbers, giving the expansion

$$\zeta(2n) = \frac{2^{2n-1} |B_{2n}| \pi^{2n}}{(2n)!}.$$

(The Bernoulli numbers may be defined by the identity

$$\frac{x}{e^x - 1} = \sum_{n=0}^{\infty} \frac{B_n x^n}{n!}.$$

Riemann at work

The first few are $B_0 = 1$, $B_1 = -1/2$, $B_2 = 1/6$, $B_4 = -1/30$, $B_6 = 1/42$, $B_8 = -1/30$, $B_{10} = 5/66$, $B_{12} = -691/2730$, $B_{14} = 7/6$, $B_{16} = -3617/510$, with $B_{2n+1} = 0$ for all $n > 0$.)

Combining Euler's result with Lindemann's 1882 proof that π is transcendental effectively proves that $\zeta(2n)$ is transcendental for all $n > 0$.

As we mentioned already, the nature of the odd terms $\zeta(2n+1)$ turn out to be significantly more difficult to determine. Computations show that the first few values are

$$\zeta(3) = 1.2020569032\ldots,$$
$$\zeta(5) = 1.0369277551\ldots,$$
$$\zeta(7) = 1.0083492774\ldots,$$
$$\zeta(9) = 1.0020083928\ldots,$$

but are some or all of the numbers $\zeta(2n + 1)$ rational or irrational?

In 1979, the French mathematician Roger Apéry managed to show that $\zeta(3)$ is irrational. Not the least remarkable aspect of this feat was that it was by far the greatest result of his career, and yet he was over sixty years of age when he did it. (As a result of his important discovery, $\zeta(3)$ is sometimes called *Apéry's constant*.) But no similar results are known for other odd integer arguments.[2]

Apéry, in his proof, made use of the series expansion of $\zeta(3)$ that is part of the following sequence of rapidly convergent formulas:

$$\zeta(2) = 3 \sum_{k=1}^{\infty} \frac{1}{k^2 \binom{2k}{k}},$$

$$\zeta(3) = \frac{5}{2} \sum_{k=1}^{\infty} \frac{(-1)^{k+1}}{k^3 \binom{2k}{k}},$$

$$\zeta(4) = \frac{36}{17} \sum_{k=1}^{\infty} \frac{1}{k^4 \binom{2k}{k}}.$$

The first formula was known in the nineteenth century, the second was known to Markoff in 1890 and was discovered several times during the twentieth century, and the last one was found in the 1970s.

These three formulas led many people to conjecture that the constant

$$Q_5 = \zeta(5) \left/ \sum_{k=1}^{\infty} \frac{(-1)^{k+1}}{k^5 \binom{2k}{k}} \right.$$

is rational, or at least algebraic. However, 10,000-digit PSLQ computations have shown that if Q_5 is a zero of an integer polynomial of degree at most 25, then the Euclidean norm of the vector of coefficients must be bigger than 1.24×10^{383}. This is suggestive that Q_5 is in fact transcendental (and that so too is $\zeta(5)$).

In the late 1990s, based on the negative outcome of the PSLQ investigation, another search was carried out for a multi-term, but otherwise similar expression for $\zeta(5)$, eventually coming up with the following:

$$\zeta(5) = 2 \sum_{k=1}^{\infty} \frac{(-1)^{k+1}}{k^5 \binom{2k}{k}} - \frac{5}{2} \sum_{k=1}^{\infty} \frac{(-1)^{k+1}}{k^3 \binom{2k}{k}} \sum_{j=1}^{k-1} \frac{1}{j^2}$$

---

[2]While it is strongly believed that all odd zeta values are irrational (in part on the basis of PSLQ-like computations), all that has been *proved* to date is that at least one of the next four odd values is irrational, as well as infinitely many other values.

together with similar expressions for $\zeta(7)$, $\zeta(9)$, and $\zeta(11)$.[3] In particular, especially striking formulas were found for $\zeta(4n+3)$. (This is described in the last example of Chapter 6.)

Turning back to the behavior of $\zeta(n)$ for even arguments, the book *Experimental Mathematics in Action* describes (Section 9.6) in some detail the experimental process that led to the discovery of a generating function for the even ζ-values.

An *ordinary generating function* for a sequence $\{a_n\}$ is a formal power series $\sum_{n=0}^{\infty} a_n x^n$. In nice cases, the sum can be evaluated in closed form and leads to a great deal of information about the sequence. For example, the generating function for the averaged harmonic sum

$$\sum_{n=1}^{\infty} \left(\sum_{k=1}^{n-1} \frac{1}{k} \right) \frac{x^n}{n}$$

evaluates to

$$\frac{1}{2} \log(1-x)^2 = \frac{1}{2}x^2 + \frac{1}{2}x^3 + \frac{11}{24}x^4 + O\left(x^5\right).$$

The main result discovered regarding the even ζ-values is the identity

$$\sum_{k=1}^{\infty} \frac{1}{k^2 - x^2} = 3 \sum_{k=1}^{\infty} \frac{1}{k^2 \binom{2k}{k} \left(1 - x^2/k^2\right)} \prod_{m=1}^{k-1} \left(\frac{1 - 4x^2/m^2}{1 - x^2/m^2} \right). \qquad (4.1)$$

The left-hand side of this identity is equal to

$$\sum_{n=0}^{\infty} \zeta(2n+2)x^{2n} = \frac{1 - \pi x \cot(\pi x)}{2x^2},$$

and so (4.1) generates an Apéry-like formula for $\zeta(2n)$ for every positive integer n. The first two specific instances are

$$\zeta(2) = 3 \sum_{k=1}^{\infty} \frac{1}{\binom{2k}{k} k^2},$$

$$\zeta(4) = 3 \sum_{k=1}^{\infty} \frac{1}{\binom{2k}{k} k^4} - 9 \sum_{k=1}^{\infty} \frac{\sum_{j=1}^{k-1} j^{-2}}{\binom{2k}{k} k^2}.$$

[3]The $\zeta(5)$ result was known to Max Koecher [Koecher 80].

Formula (4.1) turns out to be equivalent to the hypergeometric identity

$$
{}_3F_2\left(\begin{matrix} 3k, \, -k, \, k+1 \\ 2k+1, \, k+\frac{1}{2} \end{matrix} \middle| \frac{1}{4}\right) = \frac{\binom{2k}{k}}{\binom{3k}{k}}. \tag{4.2}
$$

Hypergeometric functions were first explored by Gauss, and provide a wonderful and systematic way of describing many of the special functions of mathematical physics and classical mathematics, either directly or in the limit. For example, e^x is a very special degenerate case of a hypergeometric function, as are most of everyone's favorite power series.

The definition of our ${}_3F_2$ is

$$
{}_3F_2\left(\begin{matrix} a,b,c \\ d,e \end{matrix} \middle| z\right) = \sum_{n=0}^{\infty} \frac{(a;n)(b;n)(c;n)}{(d;n)(e;n)} \frac{z^n}{n!} \tag{4.3}
$$

where $(a;n) = (a)(a+1)\ldots(a+n-1)$ is the rising factorial or Pochhammer symbol; hence $(1;n) = n!$. The ${}_2F_1$ we met in the BBP formula in Chapter 2 is the special case with $c = e$, which can be canceled.

The equivalence of (4.1) and (4.2) was first established using a computational method of Wilf and Zeilberger but was subsequently proved by human brain power alone.

Early in 2007, Neil Calkin observed that the same hypergeometric function appeared to have structure when evaluated at 1. It is unusual for such a function to have closed forms at both 1/4 and 1. The values Calkin obtained were, wait for it:

$$
1, \, -\frac{1}{3}, \, \frac{1}{25}, \, -\frac{1}{147}, \, \frac{1}{1089}, \, -\frac{3}{20449}, \, \frac{1}{48841}, \, -\frac{1}{312987}, \, \frac{25}{55190041},
$$
$$
-\frac{1}{14322675}, \, \frac{1}{100100025}, \, -\frac{49}{32065374675}, \, \ldots
$$

So now you know where that sequence in the previous chapter came from. Indeed it is known (and can be proved—see [Borwein and Bailey 08]) that, for each positive integer k, you have

$$
\sqrt{{}_3F_2\left(\begin{matrix} 6k, \, -2k, \, 2k+1 \\ 4k+1, \, 2k+\frac{1}{2} \end{matrix} \middle| 1\right)} = \frac{\binom{6k}{k}}{\binom{4k}{k}}.
$$

along with a similar formula for odd k.

Explorations

1. *Closed forms for $\zeta(2n)$, $\beta(2n+1)$.* Euler evaluated the Riemann zeta function at positive integers in a heuristic tour de force, using his discovery of the product formula for the sinc function

$$\frac{\sin(\pi x)}{\pi x} = \prod_{n=1}^{\infty} \frac{1 - x^2}{n^2},$$

which he argued should look like a "big polynomial" and be determined by its zeroes (at all the integers) and its value at zero.[4] He then equated this to the Maclaurin series and obtained the evaluations

$$\zeta(2) = \frac{\pi^2}{6}, \quad \zeta(4) = \frac{\pi^4}{90}, \quad \zeta(6) = \frac{\pi^6}{945}, \quad \zeta(8) = \frac{\pi^8}{9450}, \quad \cdots$$

In general it follows that the value of $\zeta(2n)$ will be a rational multiple of π^{2n}.

(a) Try to work out the closed form.

(b) Try to correspondingly compute the odd values of the *Catalan zeta function*

$$\beta(n) := \sum_{k=0}^{\infty} \frac{(-1)^k}{(2k+1)^n}$$

by using an appropriate product for $\cos(\pi x/2)$. Clearly, $\beta(1) = \pi/4$ and $\beta(2) = G$ (which has no known closed form), and it turns out that $\beta(3) = \pi^3/32$ and $\beta(5) = 5\pi^5/1356$.

2. *Multi-dimensional zeta functions.* Euler was also the first person to seriously grapple with multi-dimensional analogues of the ζ-function. A beautiful foundational result is

$$\zeta(2,1) := \sum_{n=1}^{\infty} \frac{1 + 1/2 + \ldots + 1/n}{(n+1)^2} = \zeta(3).$$

There are many proofs of this identity, some elementary, others less so. You are challenged to find at least one, after confirming numerically that you believe the assertion.

---

[4]This is the heuristic part, since not every analytic function has such a simple product expansion. Great mathematicians have a habit of making correct leaps of faith.

3. *The Riemann hypothesis.* The truth of the Riemann hypothesis (RH) remains unsettled. There are some eminent mathematicians who do not believe it is true and some who believe it will be resolved within a few years. J. E. Littlewood famously proved a theorem with two cases: RH holds and RH fails. (A strategy that will not satisfy an intuitionist.) While a great deal of computation has been done confirming that the non-trivial zeroes do lie on the critical line ($\text{Re}(z) = \frac{1}{2}$), there is a general consensus that the evidence so far is too little to rely on. This emphasizes one of the conundrums of experimental mathematics: what may be an overwhelming amount of evidence in some settings may be quite underwhelming in others. Only well-honed intuition, based on careful heuristic arguments and substantial knowledge, can help distinguish the two. Moreover, the RH has many equivalent reformulations, some of which seem temptingly accessible and some impossibly hard [Borwein P. et al. 07].

(a) Compute the first half-dozen zeroes of $t \mapsto \left| \zeta \left(\frac{1}{2} + it \right) \right|$ for $t > 0$ and then plot the function on the interval $[0, 40]$.

(b) Plot $(x, y) \mapsto |\zeta (x + iy)|$ for $0 < x < 1$, $1 < y < 40$. Examine the behavior of the x-cross sections.

Monsieur, où est
l'intégrale dans
votre théorie?

Sire, je n'ai pas
besoin de cette
hypothèse!

Chapter 5

⤳

Evaluate the Following Integral

Nature laughs at the difficulties of integration.
—Pierre-Simon Laplace (1749–1827)

Anyone who has taken a calculus course at high school or college has read the instruction: "Evaluate the following integral." For many students, the words fill them with dread, for others they bring a shiver of excited antici-pation. For both groups, the reason is the same: integration is hard. As an inverse operation, it requires a great deal of pattern-recognition skill and experience. Students who love a hard intellectual challenge generally find integration extremely satisfying, especially when a seemingly impossible integral turns out to have an elegant solution.

But there are integrals that even the most brilliant mathematical mind will find intractable, and then it can be profitable to call on the aid of a computer.

These days, the symbolic processors in Mathematica and Maple can handle pretty well any definite integral that has a simple enough solu-tion. Moreover, there is a decision procedure for indefinite integration, called the Risch algorithm, which is implemented to a fair degree in both packages. (It will, however, often give an answer that would not satisfy a human, as in our first example below.)

In many cases where a definite integral arises in the real world, there is no closed-form solution to the underlying indefinite integral, and nu-merical methods must be used—something again that Mathematica and especially Maple handle well.

But the technology has not turned integration into mere key pushing. Although many integrals that would be impossible to solve by hand can indeed be dealt with in a few routine keystrokes, there are many cases where it takes a genuine human-machine collaboration to complete the task. Such cases can require all the intellectual skills that hand-integration does, and as a result yield the same rewards on success. Those are the kinds of examples we'll look at here.

In engineering or physics, it is common to evaluate definite integrals numerically. The number is, after all, what the engineer or physicist often needs. In experimental mathematics, however, we sometimes evaluate an integral numerically *in order to find a closed-form solution*.

For example, using Mathematica or Maple we can evaluate the following definite integral to 100-digit accuracy:

$$\int_0^1 \frac{t^2 \log t}{(t^2 - 1)(t^4 + 1)} dt =$$

0.18067126259065494279230812898167161533711457101829676

62662407942937585662241330017708982541504837997707740...

Both packages succeed in finding an analytic integral and providing a closed form for this number, but in each case the answer is very complicated.[1] An alternative approach is to use the ISC to identify the numerical answer. It yields the elegant result

$$\int_0^1 \frac{t^2 \log t}{(t^2 - 1)(t^4 + 1)} dt = \frac{\pi^2 \left(2 - \sqrt{2}\right)}{32}.$$

A similar approach evaluates the following two integrals:

$$\int_0^\pi \frac{x \sin x}{1 + \cos^2 x} dx =$$

2.46740110027233965470862274996903778382842485181019766603337344055011205604801310750443350929638057956006...

$$= \frac{\pi^2}{4},$$

$$\int_0^{\pi/4} \frac{t^2}{\sin^2 t} dt =$$

0.84351184168503463400262005199952815165168908642144429

3697112596906587355669239938399327915596371348023976...

$$= -\frac{\pi^2}{16} + \frac{\pi \log 2}{4} + G,$$

where G is Catalan's constant

$$G = \sum_{n=0}^{\infty} \frac{(-1)^n}{(2n+1)^2},$$

which we met earlier.

Once you see these forms, it's a fairly straightforward task to evaluate the integrals analytically using either the method of residues or by Fourier techniques.[2]

Incidentally, Catalan's constant is widely believed to be irrational, but this has never been proved. Using an integer relation tool, together with a high-precision numerical value of the constant (which can easily be found by typing N[Catalan,100] in Mathematica or evalf[100](Catalan) in Maple), you can see that it is not a root of an integer polynomial with reasonable degree and reasonable-sized coefficients.

The same general approach of numerical evaluation followed by a call to the ISC also yielded the following results, which were subsequently verified analytically. Let

$$C(a) = \int_0^1 \frac{\arctan\left(\sqrt{x^2 + a^2}\right)}{\sqrt{x^2 + a^2}\,(x^2 + 1)} dx.$$

Then,

$$C(0) = \frac{\pi \log 2}{8} + \frac{G}{2},$$

---

[2]Mathematica 6.0 can now do two of these three out-of-the-box, but earlier versions were less successful. Any book like this inevitably spurs enhancements in the computer algebra systems that quickly belie authors' claims.

$$C(1) = \frac{\pi}{4} - \frac{\pi\sqrt{2}}{2} + \frac{3\sqrt{2}\arctan\left(\sqrt{2}\right)}{2},$$

$$C\left(\sqrt{2}\right) = \frac{5\pi^2}{96},$$

where again G is Catalan's constant.

Physicists often find themselves faced with particularly challenging integrals. For example, Borwein collaborated with the British physicist David Broadhurst to tackle the following nasty-looking beast:

$$I = \frac{24}{7\sqrt{7}} \int_{\pi/3}^{\pi/2} \log\left|\frac{\tan t + \sqrt{7}}{\tan t - \sqrt{7}}\right| dt,$$

which arose in quantum physics.

With the assistance of David Bailey, Borwein and Broadhurst were able to confirm that, to the 20,000-digit accuracy of the numerical evaluation, which exhibited 19,995 digits of equality,[3]

$$I = \sum_{n=0}^{\infty}\left[\frac{1}{(7n+1)^2} + \frac{1}{(7n+2)^2} - \frac{1}{(7n+3)^2} + \frac{1}{(7n+4)^2} - \frac{1}{(7n+5)^2} - \frac{1}{(7n+6)^2}\right].$$

They did not then know the identity had already been proved analytically by Don Zagier. They knew it was certainly true, as were various more complex relatives. The series is the value of a zeta-like function at 2. The evaluation was performed on the full 1024 processors of the Virginia Tech Apple G5 Terascale Computing Facility cluster in 46.15 minutes.

Given sufficient computing power, say like the Virginia Tech facility just mentioned, the same overall approach can be used to evaluate double or triple integrals. Before calling upon the heavy weaponry of parallel supercomputers, however, it sometimes pays to see what one of the "3Ms" (Maple, Mathematica, or Matlab) will do easily on a desktop computer.

---

[3]Note that $\log_{10}(20000) = 4.30\ldots$, and that you should anticipate a logarithmic round-off error in such cases. That said, anything below a 20-digit loss is more than a little impressive; and crucially, we did not tell you the working precision was 20,014 digits.

For instance, consider Borwein, Bailey, and Crandall's discovery of the following "box integral" evaluation:

$$C = \int_{-1}^{1} \int_{-1}^{1} \frac{dxdy}{\sqrt{1 + x^2 + y^2}} = 4 \log \left(2 + \sqrt{3}\right) - \frac{2\pi}{3}.$$

The group began by noting that, because of the symmetry in the integrand, computing

$$C = 4 \int_{0}^{1} \int_{0}^{1} \frac{dxdy}{\sqrt{1 + x^2 + y^2}}$$

would be quicker. (Another option, polar substitution, makes the domain less pleasant and so was not pursued initially.)

Then experience played a role. Related integrals that the team had tackled had involved homogeneous combinations of

$$a = \log \left(1 + \sqrt{3}\right), \quad b = \log 2, \quad c = \pi.$$

For example, $a^2, b^2, c^2, ab, bc, ca$ is the full set of order 2.

Looking for a linear relation between C and a, b, c of just the first order returned $[3, 24, -12, -2]$, which meant that

$$C = 8 \log \left(1 + \sqrt{3}\right) - 4 \log 2 - \frac{2\pi}{3}.$$

The hunt was done with about 12 digits and quickly confirmed to 20. This simplified to the form written above. (Since $(1 + \sqrt{3})/2$ is a unit in $Q(\sqrt{3})$, the team might well have started with its log rather than the two above. If the linear form had failed to return an answer, they could have computed C to many more digits (for example, 35) and tried the quadratic basis above or the smaller one using only π and $\log(2 + \sqrt{3})$. They didn't, but they could have.)

Another satisfying episode in the experimental evaluation of (double) integrals began quite recently with the publication of the February 2007 issue of the *American Mathematical Monthly*. One of problems published in the regular "Problems" section was to evaluate the iterated integral

$$C = \int_{0}^{\infty} \int_{y}^{\infty} \frac{(x - y)^2 \log \left((x + y)/(x - y)\right)}{xy \sinh(x + y)} dxdy.$$

When their copies of the *Monthly* arrived in the mail, Borwein and Bailey both recognized that this problem was amenable to experimental methods, and independently they began to work on it.

Bailey's approach was to calculate the original double integral, after making the minor substitution $u = x - y$, so that both integrals have constant limits. This produced the numerical result

$$C = 1.1532659890804730178602752931059938854511244009224435425100\ldots$$

Unfortunately, when he fed this number into the ISC, it was not able to recognize it.

Meanwhile, working with Maple, Borwein employed the simple substitution $x = ty$ to transform the integral into

$$C = \int_0^\infty \int_1^\infty \frac{y(t-1)^2 \log\left((t+1)/(t-1)\right)}{t \sinh(ty+y)}\,dt\,dy.$$

He then interchanged the order of integrals to produce the one-dimensional integral

$$C = \frac{\pi^2}{4} \int_1^\infty \frac{(t-1)^2 \left(\log(t+1) - \log(t-1)\right)}{t(t+1)^2}\,dt.$$

Substituting $t = 1/s$, this became

$$C = \frac{\pi^2}{4} \int_0^1 \frac{(s-1)^2 \left(\log(1+s) - \log(1-s)\right)}{s(1+s)^2}\,ds.$$

Now he was in business. Maple was able to numerically evaluate either form of the single integral (without the external coefficient) as $0.4674011002723397\ldots$, and was further able to recognize this constant as $\pi^2/4 - 2$ via the `identify` function. Thus, the entire integral was recognized as

$$C = \frac{\pi^4}{16} - \frac{\pi^2}{2}.$$

Now that Borwein "knew" the answer, it was a fairly simple matter, while still working in a Maple environment, to "prove" it. This was done by substituting $u = (1-s)/(1+s)$ in the third form above to yield the simple equivalent form

$$C = \frac{2\pi^2}{4} \int_0^1 \frac{u^2 \log u}{u^2 - 1}\,du,$$

which Maple was able to evaluate analytically to produce the closed-form result given above. (It's also possible to do it by hand. For instance, you can do it by using the geometric series and integrating term by term.)

It was disappointing (particularly to Bailey!) that the ISC was not able to recognize the numerical value of the original integral. Evidently, the number Bailey obtained lies just outside the search space and stored values that it works with. Over the summer of 2007, a revised and enhanced parallel utility came online at http://ddrive.cs.dal.ca/~isc that is able to produce the closed-form evaluation for this problem, using the original numerical value as input. Indeed, the single Maple instruction

```
identify(1.1532659890804730178602 7, BasisSizePoly=7);
```

immediately returns the same closed form that Borwein found.

As we observed earlier, physics can throw up some challenging integrals. The following comes from the area known as Ising Theory:

$$E_n = 2 \int_0^1 \cdots \int_0^1 \left(\prod_{n \geq k > j \geq 1} \frac{u_k/u_j - 1}{u_k/u_j + 1} \right)^2 dt_2 ... dt_n,$$

where $u_k = \prod_{i=1}^k t_i$.

A long series of computations and manipulations with E_5 carried out by Borwein, Bailey, and Crandall led them to the conjecture:

$$E_5 = 42 - 1984 \, Li_4 \left(\frac{1}{2} \right) + \frac{189\pi^4}{10} - 74\,\zeta(3)$$

$$- 1272 \, \zeta(3) \log 2 + 40\pi^2 \log^2 2 - \frac{62\pi^2}{3}$$

$$+ \frac{40\pi^2 \log 2}{3} + 88 \log^4 2 + 464 \log^2 2 - 40 \log 2,$$

where $Li_n(x)$ is the nth order polylogarithm

$$Li_n(x) = \sum_{k=1}^{\infty} \frac{x^k}{k^n}.$$

It's hard to imagine such an expression ever being found by an unaided human brain!

After the integral was evaluated numerically to 240-digit precision, an integer relation algorithm using 50-digit precision discovered the closed form just given. This evaluation is at least 190 digits beyond the level that could reasonably be ascribed to numerical round-off error. So it looks likely that the evaluation is correct. But no formal proof is known. Computations like this can take as much as a day on 2^9 processors. This is more than a year of time on a good desktop computer, and hence they are really not practicable without access to parallel machines.

Recently, Craig Tracy asked Borwein for help finding a closed form for a more complicated relative of E_n, D_5, where

$$D_n = \frac{4}{n!} \int_0^\infty \cdots \int_0^\infty \frac{\prod_{i<j} \left(\frac{u_i - u_j}{u_i + u_j}\right)^2}{\left(\sum_{j=1}^n u_j + 1/u_j\right)^2} \frac{du_1}{u_1} \cdots \frac{du_n}{u_n}.$$

Values for D_1, D_2, D_3, and D_4 have been known for 30 years:

$$D_1 = 2,$$

$$D_2 = \frac{1}{3},$$

$$D_3 = 8 + \frac{4\pi^2}{3} - 27L_{-3}(2),$$

$$D_4 = \frac{4\pi^2}{9} - \frac{1}{6} - \frac{7\zeta(3)}{2},$$

where $L_{-3}(2)$ is the Dirichlet series

$$L_{-3}(2) = \sum_{n=0}^\infty \left[\frac{1}{(3n+1)^2} - \frac{1}{(3n+2)^2} \right].$$

To solve Tracy's problem using the integer relation algorithm PSLQ entails being able to evaluate a five-dimensional integral to at least 50 or 250 places, so that you can search for combinations of 6 to 15 constants. Monte Carlo methods, which are random sampling methods good for finding a few digits of multi-dimensional integrals, even in high dimensions, certainly cannot do this.

Bailey, Borwein, and Crandall were able to reduce D_5 to a horrifying, several-page-long three-dimensional symbolic integral! A 256-CPU computation at Lawrence Berkeley National Laboratory provided 500 digits in 18.2 hours on *Bassi*, an IBM Power5 system. Here are those 500 digits:

0.0024846057623403154799505091539097496350606776424875161587076
9216182213785691543575379268994872451201870687211063925205511862
0699449975422656562646708538284124500116682230004545703268769 73
8489615198247961303552525851510715438638113696174922429855780 76
2804289477702787109211981116063406312541360385984019828078640 18
6930726810988548230378878848758305835125785523641996948691463 14
0911273630946052409340088716283870643642186120450902997335663 41
1372761220240883454631501711354084419784092245668504608184468...

While no closed form was found, the complete data set is available to any researcher who has a new idea and wants to continue the hunt. The dataset includes details of the basis space hunted over, and records the negative result in a fashion that makes it useful to others, exactly as would be expected in the (other) experimental sciences.

As this episode shows, integer relation methods typically run very rapidly, but there are highly taxing exceptions.

Explorations

1. *Integration*. Here are seven integrals to attempt to compute numerically and then recognize. Your computer algebra system (CAS) may be able to do them directly. In each case the answer is a combination—sums of products—of constants like e, $\sqrt{2}$, $\sqrt{3}$, π, $\zeta(3)$, $\log 2$, G, and γ, where again G is Catalan's constant and γ is Euler's gamma constant. They come from a standard table of integrals and series.

(a) $int_0^\infty \dfrac{x^2}{\sqrt{e^x - 1}} dx$

(b) $\displaystyle\int_0^{\pi/4} x \tan(x) dx$

(c) $\displaystyle\int_0^{\pi/4} (\pi/4 - x \tan x) \tan(x) dx$

(d) $\displaystyle\int_0^{\pi/2} \dfrac{x^2}{\sin^2(x)} dx$

(e) $\int_0^\infty \frac{\log(x)}{\cosh^2(x)} dx$

(f) $\int_0^{\pi/2} \sqrt{\tan x}\, dx$

(g) $\int_0^{\pi/2} \frac{x^4}{\sin^4(x)} dx$

At least one also involves $\log(\pi)$.

2. *Failure of Fubini's theorem.* Working heuristically in a computer algebra system, you can blithely take limits, interchange order of integration or summation, and so on. The latter two operations are fully justifiable when the integrand or summand is positive (via absolute convergence), but it is instructive to be reminded of what can go wrong and to explore it in a CAS. (A corresponding caution for limits is discussed in the Explorations in Chapter 9.) Determine and examine why

$$\int_0^1 \int_0^1 \frac{x^2 - y^2}{(x^2 + y^2)^2} dx dy = -\frac{\pi}{4}$$

but

$$\int_0^1 \int_0^1 \frac{x^2 - y^2}{(x^2 + y^2)^2} dy dx = \frac{\pi}{4}.$$

3. *"Impossible" integrals from physics.* Here are three integrals that look very much like those in Exploration 1 above. They arise in mathematical physics with box integrals, but no closed form is known for any of them.

(a) $\int_0^1 \frac{\log(\sqrt{3 + y^2} + 1) - \log(\sqrt{3 + y^2} - 1)}{1 + y^2} dy$

(b) $\int_3^4 \frac{\arcsec(x)}{\sqrt{x^2 - 4x - 3}} dx$

(c) $\int_0^{\pi/4} \int_0^{\pi/4} \sqrt{\sec^2(a) + \sec^2(b)}\, da\, db$

4. *Stumping Mathematica and Maple.* CASs are continually being improved. Mathematica 6 and Maple 11 (the versions current at the time of writing) are unable to evaluate the integral

$$\int_0^{\pi/2} \frac{\arcsin\left(\sin(x)/\sqrt{2}\right) \sin(x)}{\sqrt{4 - 2\sin^2(x)}}\,dx = -\frac{\pi}{8}\sqrt{2}\ln(2)$$

but ICS 2.0 with `identify` will find it.

Thomas Jefferson
signing a check
for 60 000 000 francs
to Napoleon —
the best real estate
deal in history!

Chapter 6

⤳

Serendipity

*I'm a great believer in luck, and I find the harder I work
the more I have of it.*
—Thomas Jefferson (1743–1826)

A popular portrayal of scientists in television series and movies has the scientist mixing random colored liquids and occasionally striking it lucky, discovering a potion that makes things invisible—or being unlucky and causing an explosion that makes their hair stand on end and takes off their eyebrows. Professional experimental science isn't like that, of course. (We hope you agree with that "of course.") In real life, scientists begin by formulating a hypothesis and performing an experiment to test it. But that doesn't mean that there isn't some luck involved. In fact, sometimes, a major discovery depends on an incredible stroke of luck.

A famous illustration is the discovery of penicillin by Sir Alexander Fleming in 1928. This example shows that the discovery may depend on a stroke of luck, but, as US President Thomas Jefferson indicated (see chapter quote above), it's not "blind luck." The person making the discovery has to be anything but blind, recognizing the chance occurrence as potentially important and then determining its significance. After all, Fleming might have simply mumbled "Yuck!" and then thrown away the mold-infested cultures that greeted him when he returned to the laboratory the next morning, without noticing that the mold seemed to have inhibited the growth of the bacteria he was trying to cultivate. But he didn't.[1]

The same is true in experimental mathematics. Because it *is* experimental, this approach to mathematical research can sometimes lead to serendipitous discovery. For example, computation of a formula to 20 or

---

[1] Pasteur is reported to have said that fate or chance favors the prepared mind. Nowhere is that more true than in mathematics.

30 decimal places, or more, may yield a pattern that suggests something interesting—and unsuspected—is going on. Or, if two different expressions turn out to have numerical expansions that agree to as few as five or six decimal places, then most mathematicians would agree it was worth investigating to see if the two were in fact identically equal.

This is exactly what happened in 1993, when an undergraduate at the University of Waterloo in Canada, Enrico Au-Yeung, brought to the attention of one of us (Borwein) the curious result

$$\sum_{k=1}^{\infty} \left(1 + \frac{1}{2} + \ldots + \frac{1}{k}\right)^2 k^{-2} = 4.59987$$

$$\approx \frac{17}{4}\zeta(4) = \frac{17\pi^4}{360}. \tag{6.1}$$

Au-Yeung had computed the sum in (6.1) to 500,000 terms, giving an accuracy of five or six decimal digits. Suspecting that his discovery was merely a numerical coincidence, we set out to compute the sum to a higher level of precision. Using Fourier analysis and Parseval's equation, we obtained

$$\frac{1}{2\pi}\int_0^{\pi} (\pi - t)^2 \log^2\left(2\sin\frac{t}{2}\right) dt = \sum_{n=1}^{\infty} \frac{\left(\sum_{k=1}^{n} 1/k\right)^2}{(n+1)^2}. \tag{6.2}$$

Our idea was to use the series on the right of (6.2) to evaluate (6.1), while the integral on the left could be computed using the numerical quadrature facility of Mathematica or Maple. When we did this, however, we were surprised to find that the conjectured identity holds to more than 30 digits.

What we did not know at the time was that Au-Yeung's suspected identity follows directly from a related result proved by the Dutch mathematician P. J. de Doelder in 1991. In fact, it had cropped up even earlier as a problem in the *American Mathematical Monthly*, but the story goes back further still. A bit of historical research revealed that Euler considered these summations. In response to a letter from Goldbach, he examined sums that are equivalent to

$$\sum_{k=1}^{\infty} \left(1 + \frac{1}{2^m} + \ldots + \frac{1}{k^m}\right)(k+1)^{-n}. \tag{6.3}$$

The great Swiss mathematician was able to give explicit values for certain of these sums in terms of the Riemann zeta function. For example, he found an explicit formula for the case $m = 1, n \geq 2$.

In retrospect, perhaps it was for the better that we had not known of de Doelder's and Euler's results, because Au-Yeung's intriguing numerical discovery launched a fruitful line of research by a number of researchers that continued until nearly the present day. Sums of this general form are nowadays known as "Euler sums" or "Euler-Zagier sums."

In order to explore these sums more rigorously, we found it necessary to develop an efficient means to calculate their value to high precision, specifically the 200 or more digit accuracy needed to obtain numerically significant results using integer relation calculations. High-precision calculations of many of these sums, together with considerable investigations involving heavy use of Maple's symbolic manipulation facilities, eventually yielded numerous new results.

Here are just two of the many interesting results that we first discovered numerically, which have since been established analytically:

$$\sum_{k=1}^{\infty} \left(1 + \frac{1}{2} + \ldots + \frac{1}{k}\right)(k+1)^{-4} = \frac{37}{22680}\pi^6 - \zeta^2(3),$$

$$\sum_{k=1}^{\infty} \left(1 + \frac{1}{2} + \ldots + \frac{1}{k}\right)^3 (k+1)^{-6} =$$

$$\zeta^3(3) + \frac{197}{24}\zeta(9) + \frac{1}{2}\pi^2\zeta(7) - \frac{11}{120}\pi^4\zeta(5) - \frac{37}{7560}\pi^6\zeta(3).$$

Since these results were first obtained in 1994, many more specific identities have been discovered, and a growing body of general formulas and other results have been proved.

Another occasion when researchers got lucky occurred in the late sixties, when Ronald Graham and Henry Pollak were studying the sequences (a_n) and (b_n) defined by starting with $a_0 = m$ and then iterating

$$a_{n+1} = \left\lfloor \sqrt{2a_n(a_n + 1)} \right\rfloor,$$

where $\lfloor \ldots \rfloor$ denotes integer part, as usual, and then defining

$$b_n = a_{2n+1} - 2a_{2n-1}.$$

"Why would they look at such sequences?" you ask. They arose out of an investigation of sorting algorithms, that's why. In any event, Graham and Pollak wondered if they could identify either or both of these sequences. After playing around with the sequences for some time, they found that if you define the constants $\alpha(m)$ for integers $m \geq 1$ by $\alpha(m) = 0.b_1 b_2 b_3 \ldots_2$, where the final subscript 2 means that the sequence (b_n) is to be interpreted as the binary expansion of the constant $\alpha(m)$, then the resulting constants are simple algebraic numbers. In particular,

$$
\begin{aligned}
\alpha(1) &= \sqrt{2} - 1, & \alpha(2) &= \sqrt{2} - 1, \\
\alpha(3) &= 2\sqrt{2} - 2, & \alpha(4) &= 2\sqrt{2} - 2, \\
\alpha(5) &= 3\sqrt{2} - 4, & \alpha(6) &= 4\sqrt{2} - 5, \\
\alpha(7) &= 3\sqrt{2} - 4, & \alpha(8) &= 5\sqrt{2} - 7, \\
\alpha(9) &= 4\sqrt{2} - 5, & \alpha(10) &= 6\sqrt{2} - 8.
\end{aligned}
$$

This recognition led to an explicit formula for the sequence (a_n) as

$$
a_n = \left\lfloor \tau_m \left(2^{(n-1)/2} + 2^{(n-2)/2} \right) \right\rfloor,
$$

where τ_m is the mth smallest real number in the set

$$
\{1, 2, 3, \ldots\} \cup \left\{ \sqrt{2}, 2\sqrt{2}, 3\sqrt{2}, \ldots \right\}.
$$

It is not known if there are analogous properties for generalized sequences such as

$$
a_{n+1} = \left\lfloor \sqrt{3 a_n (a_n + 1)} \right\rfloor,
$$

$$
a_{n+1} = \left\lfloor \sqrt[3]{2 a_n (a_n + 1)(a_n + 2)} \right\rfloor.
$$

For our final example of "proof by serendipity," we begin with a discovery we made together with D. M. Bradley in 1996. We were looking for infinite-series (Apéry-like) formulas for integer values of the Riemann zeta function, similar to those described in Chapter 4.

After hours of computer time running integer relation algorithms, we were able to come up with the following possible identity, seemingly valid for any complex number z such that $|z| < 1$:

$$\sum_{n=0}^{\infty} \zeta(4n+3)z^{4n} = \sum_{k=1}^{\infty} \frac{1}{k^3(1-z^4/k^4)}$$

$$= \frac{5}{2} \sum_{k=1}^{\infty} \frac{(-1)^{k-1}}{k^3\binom{2k}{k}(1-z^4/k^4)} \prod_{m=1}^{k-1} \frac{1+4z^4/m^4}{1-z^4/m^4}.$$

For $z = 0$, the identity gives Apéry's formula for $\zeta(3)$, which we encountered in Chapter 4.

The computer data told us that this identity was numerically valid for all values of n from 1 to 10. But could we prove the identity analytically? The equality on the top line did not cause us any difficulty, but the second one had a totally unexpected form. The computer work pointed to a possible proof.

What the computations for the first ten cases actually showed was that

$$\sum_{n=0}^{\infty} \zeta(4n+3)z^{4n} = \sum_{k=1}^{\infty} \frac{1}{k^3(1-z^4/k^4)}$$

has the form

$$\frac{5}{2} \sum_{k\geq 1} \frac{(-1)^{k-1}}{k^3\binom{2k}{k}} \frac{P_k(z)}{(1-z^4/k^4)}$$

for *some* formula P_k of z. But we had enough computer-generated data to compute a closed form for P_k, namely,

$$P_k(z) = \prod_{m=1}^{k-1} \frac{1+4z^4/m^4}{1-z^4/m^4}.$$

After a further week's work, we came up with the following intriguing finite-sum identity, equivalent to our original one:

$$\sum_{k=1}^{n} \frac{2n^2}{k^2} \frac{\prod_{i=1}^{n-1}(4k^4+i^4)}{\prod_{i=1,i\neq k}^{n-1}(k^4-i^4)} = \binom{2n}{n}.$$

This version was not proved until a year or so later, when Gert Almkvist and Andrew Granville managed to knock it off, thereby completing the proof of the original conjectured identity. Its initial discovery came about as a result of a typo in entering data at the computer keyboard! This was a serendipitous moment in the story.

In typing a formula into Maple, we mistakenly typed `infty` (the TeX command to produce the ∞ symbol) instead of Maple's reserved word `infinity`. What a surprise when Maple, which took `infty` to be a name for an integer, returned an answer. Clearly, the program knew a method for handling such finite sums. (Indeed, it knew about R. W. Gosper's work on "creative telescoping" and this is what led to the second, ultimately provable form.)

The story has a pleasing end. We showed the finite sum result to Paul Erdős shortly before his death. The famous Hungarian mathematician rushed off to think about it and returned half an hour later saying that he had no idea how to prove it, but that if it were true, it had implications for Apéry's result! Unbeknownst to Erdős, that was where the entire story began.

Serendipity can also bring to light an error in earlier work that might otherwise remain forever (or at least for a long time) undetected. David Bailey tells the following story:

> In the course of our research, Jon [Borwein] and I have encountered situations where we discovered mathematical errors, either in other work or even in our own, in the course of doing computations. In some cases we were deliberately double-checking results at the time, as part of preparing a manuscript for final publication, but more often we computationally stumbled on a mistake. I sometimes think that this application of computational techniques arguably represents the most valuable "practical" contribution of experimental mathematics. Here is an example, which arose just a few days ago. I was in the laborious process of making the very numerous corrections specified by the copyeditor to a 110-page supplement that will be included in the second edition of *Mathematics by Experiment*. At one point in our manuscript, Jon and I deduced and presented these two identities:

$$_3F_2\left(\begin{matrix} 6k,-2k,2k+1 \\ 4k+1,2k+\frac{1}{2} \end{matrix}\middle| 1\right) = \frac{(4k)!(3k)!}{((6k)!k!)^2},$$

$$_3F_2\left(\begin{matrix} 6k+3,-2k-1,2k+1 \\ 4k+3,2k+\frac{3}{2} \end{matrix}\middle| 1\right) = -\frac{1}{3}\frac{(4k+1)!(3k)!}{((6k+1)!k!)^2}$$

[as considered in Chapter 4]. I checked these two results computationally using Mathematica, evaluating the left-hand side (LHS) and right-hand side (RHS) numerically for $k = 1, 2, \ldots, 20$. The first identity checked out for all 20 values of k, but the second one did not—the LHS terms were positive while the RHS terms were negative (but absolute values were equal). I sent a note to Jon saying that there was a typo here—that most likely the minus sign on the RHS should be removed. But Jon pointed out that since these values are derived from the even-numbered elements of an alternating sequence presented a few lines earlier in our manuscript, the minus sign must stay. He then made a very nasty comment (using a word not printable here) about Mathematica's hypergeometric function facilities.

As it turned out, we were wrong and Mathematica was right. The problem was not in the RHS, but instead in the LHS—the term "$2k + 1$" should be "$2k + 2$." After making this change, everything matched. For some strange reason, the typo had the effect of flipping the signs of the LHS for all k. The bottom line is that the bug would have gone undetected, almost certainly appearing in print, if we had not checked the identities with Mathematica.

Serendipity strikes again.

Bailey's story brings to mind an episode from Douglas Adams's hilarious comedy series *The Hitchhiker's Guide to the Galaxy*, first broadcast on BBC Radio in 1978:

> The Infinite Improbability Drive is a wonderful new method of crossing vast interstellar distances in a mere nothing-th of a second without all that tedious mucking about in hyperspace.

> It was discovered by a lucky chance, and then developed into a governable form of propulsion by the Galactic Government's research team on Damogran.

> This, briefly, is the story of its discovery.

> The principle of generating small amounts of *finite* improbability by simply hooking the logic circuits of a Bambleweeny 57 sub-meson Brain to an atomic vector plotter suspended in a strong Brownian Motion producer (say a nice hot cup of tea) were of course well understood—and such generators were often used to break the ice at parties by making all the molecules in the hostess's undergarments leap simultaneously one foot to the left, in accordance with the Theory of Indeterminacy.

Many respectable physicists said that they weren't going to stand for this—partly because it was a debasement of science, but mostly because they didn't get invited to those sort of parties.

Another thing they couldn't stand was the perpetual failure they encountered in trying to construct a machine which could generate the *infinite* improbability field needed to flip a spaceship across the mind-paralyzing distances between the furthest stars, and in the end they grumpily announced that such a machine was virtually impossible.

Explorations

Webster's Dictionary defines serendipity as:

1. an aptitude for making desirable discoveries by accident.

2. good fortune; luck: *the serendipity of getting the first job she applied for.*

[1754; SERENDIP + -ITY; Horace Walpole so named a faculty possessed by the heroes of a fairy tale called *The Three Princes of Serendip*].

What can you do to increase the chance of beneficial serendipity? We don't have any exercises to assist in becoming serendipitous, but in experimental mathematics, you can cultivate certain habits of the mind that can help, such as:

○ Use tools like Sloane's encyclopedia and the ISC frequently enough that they remain in the front of your mind.

○ Try "googling" intelligently, use MathSciNet, and remember the library catalogue is still a valuable resource, as is Amazon.

○ Keep good records and try to organize and annotate code.

More mathematically:

○ Find something to plot. For example, if you are told that for $0 < x < \pi/2$,

$$2 + \frac{2}{45}x^3 \tan x > \frac{\sin^2 x}{x^2} + \frac{\tan x}{x} > 2 + \frac{16}{\pi^4}x^3 \tan x > 2$$

and that the constants are the best possible, plotting will probably reveal much more than value-checking or calculus.

○ Compute to very low precision to see if any significant precision numerical work is practicable without significant effort. For example,

$$\sum_{n=1}^{\infty} H_n^2/n^3$$

is unlikely to evaluate numerically,[2] but the sum to 100,000 places will be accurate to quite a few digits. (How many?)

○ Try the indefinite integral, the finite sum, exchanging variables and exploiting symmetries. For example, you will find it easy to identify

$$\sum_{n=0}^{\infty} \frac{(4n+3)}{(n+1)^2} \frac{\binom{2n}{n}^2}{2^{4n}}.$$

To find a proof, look at the finite sum.

○ Play with changes of variable and integration by parts in integration, partial fractions, and continued fractions.

All of these are easy on the computer and many are painful by hand. So in some real sense they merit inclusion in Chapter 8.

---

[2] As always, $H_n = 1 + 1/2 + \ldots + 1/n$.

Chapter 7

⤳

Calculating π

I am ashamed to tell you to how many figures I carried these calculations [of π] having no other business at the time.

—Isaac Newton (1642–1727)

Thousands of years ago, the ancient Greeks (and other early civilizations) noticed that if you take any circle, no matter what its size, and divide the circumference by the diameter, you will always get the same answer, a number between 3 and 4 that we now refer to as π. As a non-integer of great importance, there was always interest in calculating the most exact value of this constant—a task pursued initially for practical reasons, and then later, when the first dozen or so decimal places had been nailed down, primarily as an esoteric, academic challenge.[1]

Much of this work, although computational, does not classify as experimental mathematics.[2] But the search for ever more efficient algorithms to compute π, in particular, has occasionally made use of experimental methods, as we shall see.

Around 2,000 BC, the ancient Babylonians (implicitly) used the approximation $3^1/_8$ (3.125), while the ancient Egyptians took π to be 256/81 (3.1604...). The first mathematical determination of its value was by Archimedes around 250 BC, who used geometric reasoning to show that

---

[1]As Simon Newcomb (1835–1909) observed in the nineteenth century, "Ten decimal places of π are sufficient to give the circumference of the earth to a fraction of an inch, and thirty decimal places would give the circumference of the visible universe to a quantity imperceptible to the most powerful microscope." (Quoted in [MacHale 93].)

[2]Unlike the recent computations of binary and hexadecimal digits of π, which we described in Chapter 2. Those calculations used a formula that *was* discovered by experimental mathematics.

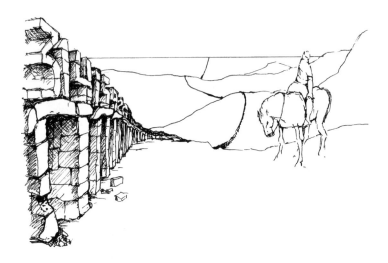

$3^{10}/71 < \pi < 3^1/7$. Archimedes effectively gave an algorithm that could be used to calculate π to any desired degree of accuracy, to wit:

$$a_0 = 2\sqrt{3}; \; b_0 = 3;$$

$$a_{n+1} = \frac{2a_n b_n}{a_n + b_n}; \; b_{n+1} = \sqrt{a_{n+1}b_n}.$$

This recursion converges to π, with the error decreasing by a factor of approximately four with each iteration. Variations of Archimedes' geometrical scheme were the basis for all high-accuracy calculations of π for the next 1,800 years. For example, in the fifth century AD, the Chinese mathematician Tsu Chung-Chih used a form of this method to compute π correct to seven digits.

With Newton and Leibniz's discovery of calculus in the seventeenth century, mathematicians had a new source of formulas for calculating π. One early calculus-based formula comes from the integral

$$\arctan x = \int_0^x \frac{dt}{1+t^2} = \int_0^x \left(1 - t^2 + t^4 - t^6 + \ldots\right) dt$$

$$= x - \frac{x^3}{3} + \frac{x^5}{5} - \frac{x^7}{7} + \frac{x^9}{9} - \ldots$$

Substituting $x = 1$ gives the so-called Gregory-Leibniz series:

$$\frac{\pi}{4} = 1 - \frac{1}{3} + \frac{1}{5} - \frac{1}{7} + \frac{1}{9} - \frac{1}{11} + \cdots$$

This series is of theoretical interest, but of no practical value in computing π, since it requires hundreds of terms to get just two decimal places of accuracy. However, by employing the trigonometric identity

$$\frac{\pi}{4} = \arctan\left(\frac{1}{2}\right) + \arctan\left(\frac{1}{3}\right)$$

(attributed to Euler in 1738), you can obtain

$$\frac{\pi}{4} = \frac{1}{2} - \frac{1}{3.2^3} + \frac{1}{5.2^5} - \frac{1}{7.2^7} + \cdots$$
$$+ \frac{1}{3} - \frac{1}{3.3^3} + \frac{1}{5.3^5} - \frac{1}{7.3^7} + \cdots,$$

which converges much more rapidly.

An even faster formula, discovered a generation earlier by John Machin, can be obtained using the identity

$$\frac{\pi}{4} = 4\arctan\left(\frac{1}{5}\right) - \arctan\left(\frac{1}{239}\right)$$

in a similar way. This formula was used in numerous computations of π, culminating with William Shanks's computation of π to 707 decimal digits in 1874. (It was later discovered that this result was in error after the 527th decimal place).

Once mathematicians had computed enough digits of π to suffice for all practical, real-world applications, one motivation to continue the hunt was very much in the spirit of modern experimental mathematics, namely, to see if the decimal expansion repeats, which would mean that π is rational. The question of the rationality of π was settled in the late 1700s, when Johann Lambert and Adrien-Marie Legendre proved it is not. A century later, in 1882, Carl Lindemann proved that it is transcendental.

The first computation of π to use a mechanical calculating device was carried out in 1945, when D. F. Ferguson computed π to 530 decimal digits, using the formula

$$\frac{\pi}{4} = 3\arctan\left(\frac{1}{4}\right) + \arctan\left(\frac{1}{20}\right) + \arctan\left(\frac{1}{1985}\right).$$

Continuing the computation over the next two years he increased this to 808 digits. In so doing, he discovered the error in Shanks's computation that we mentioned above.

A mere four years later, in 1949, under the direction of John von Neumann, the ENIAC became the first "modern" computer to calculate π, determining 2037 decimal places in 70 hours of runtime.[3]

An experimental approach can be adopted to discover other arctan formulas for π like the ones above. The idea is to simply explore for them using the numerical values of individual arctan formulas. For instance, by computing values of the individual arctans below to moderately high precision, and applying an integer relation tool, one can easily deduce the relations

$$\pi = 48 \arctan \frac{1}{49} + 128 \arctan \frac{1}{57} - 20 \arctan \frac{1}{239} + 48 \arctan \frac{1}{110443},$$

$$\pi = 176 \arctan \frac{1}{57} + 28 \arctan \frac{1}{239} - 48 \arctan \frac{1}{682} + 96 \arctan \frac{1}{12943}.$$

These particular formulas were used by Yasumasa Kanada of the University of Tokyo to compute π to a record one trillion decimal places in 2002.

Electronic computation became more efficient when π-hunters learned about the following remarkable formula of Srinivasa Ramanujan, dating from around 1910:

$$\frac{1}{\pi} = \frac{2\sqrt{2}}{9801} \sum_{k=0}^{\infty} \frac{(4k)!(1103 + 26390k)}{(k!)^4 396^{4k}}.$$

Each term in this infinite series produces an additional *eight* correct decimal places of π. In 1985, this formula was used to compute π to 17 million places.

An even more productive formula was discovered by David and Gregory Chudnovsky:

$$\frac{1}{\pi} = 12 \sum_{k=0}^{\infty} \frac{(-1)^k (6k)!(13591409 + 545140134k)}{(3k)!(k!)^3 640530^{3k+3/2}}.$$

Each term in this series produces an additional fourteen (correct) digits. The Chudnovskys used this formula to compute π to over four billion decimal places in 1994.

---

[3]Von Neumann also arranged for the computation of e.

A still more efficient method to compute π was developed independently in 1976 by Eugene Salamin and Richard Brent, taking their lead from work of Gauss a century earlier. The Salamin-Brent algorithm runs as follows: Set $a_0 = 1$, $b_0 = 1/\sqrt{2}$, and $s_0 = 1/2$, and then iterate with

$$a_k = \frac{a_{k-1} + b_{k-1}}{2}; \quad b_k = \sqrt{a_{k-1}b_{k-1}};$$

$$c_k - a_k^2 - b_k^2; \quad s_k = s_{k-1} - 2^k c_k;$$

$$p_k = \frac{2a_k^2}{s_k}.$$

Then, p_k converges *quadratically* to π. Each iteration of this algorithm approximately *doubles* the number of correct digits. Successive iterations produce 1, 4, 9, 20, 42, 85, 173, 347, and 697 correct decimal digits of π. Twenty-five iterations are sufficient to compute π to over 45 million decimal digit accuracy. (However, each of these iterations must be performed using a level of numeric precision that is at least as high as that desired for the final result.)

In the mid-1980s, Jonathan and Peter Borwein developed a number of even more productive procedures of this type, including the following:

$$a_0 = 6 - 4\sqrt{2}; \quad y_0 = \sqrt{2} - 1;$$

$$y_{k+1} = \frac{1 - (1 - y_k^4)^{1/4}}{1 + (1 - y_k^4)^{1/4}};$$

$$a_{k+1} = a_k(1 + y_{k+1})^4 - 2^{2k+3}y_{k+1}(1 + y_{k+1} + y_{k+1}^2).$$

The sequence a_k converges *quartically* to $1/\pi$. This particular algorithm, together with the Salamin-Brent scheme, was used by Kanada to compute π to over 206 billion places in 1999.

As we mentioned earlier, with the advent of the electronic computer, computation of ever more decimal digits of π rapidly became very much a game. But as often with mathematics, it was a game that had practical payoffs. Computing billions of digits of π provides an excellent way to test new computer hardware and software. The idea is to carry out the computation twice, using two different algorithms, running on two different machines with different systems, and then comparing the two results.

If the two results match place for place, you can be as close to certain as is possible in the real world that both machines are operating correctly. But if the answers differ in even one place, you know that at least one of the machines is operating incorrectly. Kanada's record-breaking π computations have all been carried out to test new supercomputer systems. (Perhaps it is more accurate to say that this is why he was given so much time on such expensive machines!)

Another payoff of record π calculations has been in the area of mathematical culture. For years, the assertion that the sequence "0123456789" appears in the decimal expansion of π was presented to students as a standard example of a simple mathematical statement that is clearly either true or false but for which we will never know which is the case. In 1997, Kanada found exactly that sequence, beginning at position 17,387,594,880. Okay, that's not going to change the world. But it sure is neat!

The digits-sequence example is actually a special case of a more general question that could turn out to be mathematically important—namely, whether π is *normal*.

A real number α is said to be normal if *every* sequence of k consecutive digits in its decimal expansion appears with limiting frequency 10^{-k}. Thus, the limiting frequency of any single digit in its decimal expansion is $1/10$, the limiting frequency of any two-digit pair is $1/100$, and so forth. In terms of measure theory, it is known that almost all real numbers are normal, yet not a single irrational algebraic number has been shown to be normal.

Examination of the expansions of π obtained by Kanada and others suggest that π is normal, but this has not been proved either. (In fact, the only provably-normal irrational numbers known are numbers specifically constructed to be normal.) This amounts to weak experimental evidence for the normality of π.

Table 1 gives statistics for the first trillion decimal digits of π.

If π were ever shown to be normal, then it would follow at once that the expansion of π contains a numeric coding of the Bible (and the Koran, and *Moby Dick*; indeed, any written text) as a consecutive sequence of digits, a development that would inevitably give rise to a distressing repeat of the wretched "Bible Code" nonsense that grabbed the headlines a few years ago, with the publication of a mass-market book of that title that claimed there were hidden messages from God in the text of the Bible.

Digit	Occurrences
0	99999485134
1	99999945664
2	100000480057
3	99999787805
4	100000357857
5	99999671008
6	99999807503
7	99999818723
8	100000791469
9	99999854780
Total	1000000000000

Table 1. Frequencies of the first trillion decimal digits of π.

Explorations

1. *Fast arithmetic.* In order to perform high-precision computation, you first have to build a high or arbitrary precision "four-function" calculator ($\pm \times \div \sqrt{}$). Only addition and subtraction are done efficiently by the familiar hand method. Multiplication of two n-digit numbers can be implemented to reduce its operational complexity from $O(n^2)$ to roughly $O(n \log n)$ by using the *fast Fourier transform* (FFT), discovered around 1963. We won't try to explain the FFT here, but we will mention that its use has revolutionized medical and geo-science computing and various other areas. You will get a taste of the injunction to "think outside the box" by observing that

$$(a+c10^n)(b+d10^n)$$

$$= a \cdot b + (a \cdot d + b \cdot c)10^n + c \cdot d10^{2n}$$

$$= a \cdot b + ((a+c)(b+d) - a \cdot b + c \cdot d)10^n + c \cdot d10^{2n}$$

can be used with moderate care to reduce a multiplication of two $2n$-digit numbers by three multiplications of n-digit numbers—at the expense of storing $a \cdot b$ and $c \cdot d$, and doing a few more additions and subtractions. This is called *Karatsuba multiplication*. It grows like $O(n^{\log_2 3})$ rather than $O(n^2)$. Since $\log_2 3 \approx 1.584962501$, this is a substantial saving, even occasionally for numbers with hundreds of digits.

Once you have implemented a fast multiplication algorithm, division of a by b reduces to multiplying a by $1/b$.

(a) Implement Newton's method to solve $b = 1/x$. Start close enough to the answer so that you are doubling precision at each step and observe that you only need to keep the first half of the result, because the second half will be over-written. Thus, the total precision used will be only twice the final precision.

(b) Likewise, solve $b = 1/x^2$.

2. *Monte Carlo calculation of π*. Monte Carlo simulation was pioneered during the Manhattan project by Stanislaw Ulam (1909–1984) and oth-

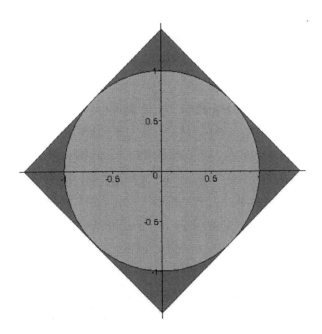

Figure 2. The 2×2 square and the unit circle used to estimate π.

ers, who recognized that this scheme permitted simulations beyond the reach of conventional methods on the limited computer systems then available. Nowadays, Monte Carlo methods are quite popular because they are well suited to parallel computation. We illustrate the Monte Carlo method with a calculation of π. It is a poor method to compute π, but well illustrative of this general class of computation.

Design and implement a Monte Carlo simulation for π, based on generating pairs of uniformly distributed numbers in the 2×2 square and testing whether they lie inside the unit circle, as shown in Figure 2.

Use, for example, the *pseudorandom* number generator $x_0 := 314159$ and $x_n := cx_{n-1} \mod 2^{32}$, where $c = 5^9 = 1953125$. This generator is of the well-known class of *linear congruential generators* and has period 2^{32}. Clearly, the probability that a pair inside the square lies in the circle is $\pi/4$, in theory if not in practice.[4]

3. *Convergence rates.* To appreciate this example, you need to work to very high precision. Let $1/\sqrt{2}$, and for $n \geq 0$ define

$$a_{n+1} := \frac{1 - \sqrt{1 - a_n^2}}{1 + \sqrt{1 - a_n^2}}, \quad \omega_n := 2^n \sqrt{4/a_{n+1}}.$$

(a) Determine the limit ω of ω_n, and examine the rate of convergence.

(b) Compare the rate of convergence of $2^n \sqrt{1/a_{n+1}}$, which also converges to ω.

---

[4]For current best-practice methods along these lines (so-called quasi-Monte Carlo methods), see [Crandall and Pomerance 05, Chapter 8].

Chapter 8

~~→

The Computer Knows More Math Than You Do

Dave: Open the pod bay doors, HAL.
HAL: I'm sorry Dave, I'm afraid I can't do that.
Dave: What's the problem?
HAL: I think you know what the problem is just as well as I do.
Dave: What are you talking about, HAL?
HAL: This mission is too important for me to allow you to jeopardize it.
Dave: I don't know what you're talking about, HAL.
—dialog from the movie *2001: A Space Odyssey*

Okay, the chapter title is a bit provocative. We admit that computers, as inanimate devices, don't actually "know" anything. (In the movie they do. At one point in *2001: A Space Odyssey*, the on-board system-control computer HAL declares, "I am putting myself to the fullest possible use, which is all I think that any conscious entity can ever hope to do." But HAL is fiction, although many of the leading artificial intelligence researchers in the late twentieth century reported that they first became interested in the subject after seeing the movie.)

But for all that real computers are not conscious, they do store a lot more information than any human, and they can generally access it a lot faster, and sometimes the astute human can take advantage of this fact.

For instance, in November 2000, the famous computer scientist Donald Knuth asked readers of the "Problems" section of the *American Mathematical Monthly* to evaluate the sum

$$S = \sum_{k=1}^{\infty} \left(\frac{k^k}{k!e^k} - \frac{1}{\sqrt{2\pi k}} \right).$$

The computer knows more than I do? We'll see about that!

One of us (Borwein) decided to try to solve it using an experimental approach.

Our first step was to compute an approximate value of S using Maple, which produced the answer (to 20 decimal places)

$$S \approx -0.08406950872765599646.$$

Feeding this numerical value to the ISC, we found ourselves looking at the following expression that the system returned:

$$S \approx -\frac{2}{3} - \frac{1}{\sqrt{2\pi}}\zeta\left(\frac{1}{2}\right),$$

where ζ is the Riemann zeta function.

Since Knuth asked for a closed-form *evaluation* of his original expression, this answered his question, but it remained open whether the two formal expressions are mathematically identical. A computer calculation quickly verified that they were equal up to 100 places and almost as quickly extended that to 500 decimal places. Thus, 16 digits of numerical data led to a prediction that was soon confirmed to compellingly many orders

By golly – she does!

of magnitude. But was the result correct? In looking for a way to prove it analytically, we had two clues to go on. The first clue was that Maple returned a high-precision numerical answer to the initial sum very quickly, despite the fact that the series converges very slowly. Obviously, the program had been doing something very clever that we weren't aware of. Upon investigation, we discovered what it was. It was using something called the *Lambert W function*:

$$W(z) = \sum_{k=1}^{\infty} \frac{(-k)^{k-1} z^k}{k!},$$

which is the inverse of $w(z) = ze^z$. We'll say a bit about this function later.

The second clue was the appearance of $\zeta(1/2)$, together with an obvious allusion in Knuth's original problem to Stirling's formula

$$\lim_{k \to \infty} \sqrt{2\pi k} \frac{k^k}{k! e^k} = 1$$

(which is presumably what led Knuth to formulate his problem).

Based on these clues, we eventually came up with the key conjecture:

$$\sum_{k=1}^{\infty} \left(\frac{1}{\sqrt{2\pi k}} - \frac{P(1/2, k-1)}{(k-1)!\sqrt{2}} \right) = \frac{1}{\sqrt{2\pi}} \zeta\left(\frac{1}{2}\right),$$

where $P(x, n) = x(x+1)\ldots(x+n-1)$ is the Pochhammer function. Maple was able to verify this conjecture symbolically, which meant that to complete the solution to Knuth's problem we needed to show that

$$\sum_{k=1}^{\infty} \left(\frac{k^k}{k!e^k} - \frac{P(1/2, k-1)}{(k-1)!\sqrt{2}} \right) = -\frac{2}{3}.$$

Remembering the relevance to the problem of the Lambert function $W(z)$ that Maple had so helpfully noted, an appeal to Abel's limit theorem suggested the possible identity

$$\lim_{z \to 1} \left(\frac{dW(-z/e)}{dz} + \frac{1}{2 - 2z} \right) = \frac{2}{3}.$$

(The binomial coefficients in the left-hand side of our key conjecture are the same as those of the function $1/\sqrt{2 - 2z}$.)

When Maple duly verified this identity, the solution was complete.

This experience was as good an illustration of human-machine collaboration as one could ever hope for, with the machine providing "insight" and "ideas," as well as number crunching and algorithmic symbol manipulation. Of course, it also required a human who was very familiar with the domain! Such a use of the computer is called "instrumental" and offers to transform how mathematicians do mathematics as this kind of investigation becomes more common.

The history of the Lambert W function is interesting. It was known to Lambert but was never given a name. The designation W was first used in 1925. In the 1990s, Gaston Gonnet, along with Rob Corless and Dave Jeffrey, introduced it into Maple. It is also implemented in Mathematica. This means that those computer packages have more or less as much "knowledge" about W as they do about exp or log, even though their users almost certainly do not. For example, typing in the Maple command

```
series(LambertW(x),x=-1...infinity)
```

tells us that its Taylor series at zero is

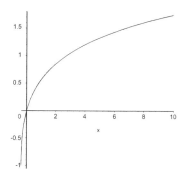
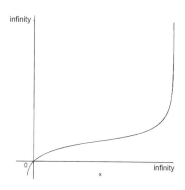

Figure 3. The Lambert W function. A correct plot (left) and a misleading one (right).

```
> 1*x-1*x^2+3/2*x^3-8/3*x^4+125/24*x^5-
54/5*x^6+16807/720*x^7-16384/315*x^8+O(x^9)
```

from which it's possible to determine the closed form for the series.

Figure 3 shows a plot of the function on $[-1, 10]$ and a misleading plot on $[-1/e, \infty)$. (Plotters often have trouble on an unbounded interval and introduce spurious inflection points.)

Although "Lambert W function" is not exactly a household term, even in mathematical circles, running a Google search on the exact phrase early in 2008 returned around 10,000 hits. When we tried, the top entry was http://mathworld.wolfram.com/LambertW-Function.html, which will tell you an enormous amount.

Another episode where the computer was able to show off its mathematical knowledge began with the publication of the January 2002 issue of *SIAM News*, the monthly newspaper produced by SIAM, the Society for Industrial and Applied Mathematics. In that issue, Nick Trefethen, of Oxford University, presented ten problems used in teaching modern graduate numerical analysis students at the university. The answer to each problem was, he said, a certain real number. Readers were challenged to compute ten digits of each answer, with a $100 prize to the best entrant.

A total of 94 teams, representing 25 countries, sent in entries. Twenty of these teams received a full 100 points (10 correct digits for each problem). At this point, an initially anonymous donor came forward to provide the money Trefethen needed to pay out a completely unexpected $2,000. (He had originally declared, "If anyone gets 50 digits in total, I will be impressed.")

Problem 9 on Tefethen's list read as follows:

The integral

$$I(\alpha) = \int_0^2 [2 + \sin(10\alpha)]x^\alpha \sin\left(\frac{\alpha}{2-x}\right) dx$$

depends on the parameter α. What is the value of $\alpha \in [0,5]$ at which $I(\alpha)$ achieves its maximum?

Like several others, we decided to tackle this problem by seeing first if Maple knew anything about the given integral. And, by golly, it did! What we did not know, and most likely you didn't either, is that the maximum parameter turns out to be expressible in terms of something called a *Meijer G function*. What's *that* you say? Well, you could do a lot worse than ask Maple, and if you do it will tell you:

```
The Meiger G function is defined by the inverse
                Laplace transform

MeigerG([as,bs],[cs,ds],z)
            /
     1      |   GAMMA(1-as+y) GAMMA(cs-y)   y
  = -----   0   --------------------------- z  dy
    2 Pi I  |   GAMMA(bs-y) GAMMA(1-ds+y)
            /
            L
where
   as = [a1,...,am],  GAMMA(1-as+y) = GAMMA(1-a1+y) ... GAMMA(1-am+y)
   bs = [b1,...,bn],  GAMMA(bs-y) = GAMMA(b1-y) ... GAMMA(bn-y)
   cs = [c1,...,cp],  GAMMA(cs-y) = GAMMA(c1-y) ... GAMMA(cp-y)
   ds = [d1,...,dq],  GAMMA(1-ds+y) = GAMMA(1-d1+y) ... GAMMA(1-dq+y)
```

The consequence of this discovery, and others like it, is that all ten of Trefethen's problems have been solved to hundreds of digits, and all except one to 10,000 decimal places. There is a lovely book by Folkmar Bornemann, Dirk Laurie, Stan Wagon, and Jörg Waldvogel, called *The SIAM 100-Digit Challenge: A Study in High-Accuracy Numerical Computing*, which describes the various solution techniques for all these problems in detail. Almost without exception, smart use of any of the 3Ms led to solutions, and in seven cases, proofs of their correctness as well! This dramatically benchmarks the changing nature of modern numerical analysis. It is no longer merely "the science of round-off errors," as it was often somewhat dismissively described at one period.

For our last example of a situation where a computer package turns out to "know" a lot more than its user, we first need to introduce the *elliptic integral* function

$$K(k) = \int_0^{\pi/2} \frac{1}{\sqrt{1 - k^2 \sin^2 \phi}} d\phi.$$

More precisely, K is a "complete elliptic integral of the first kind." Elliptic integrals get their name because they arise when you try to compute the arclength of an ellipse.

Physics students often meet elliptic integrals for the first time when they study the simple pendulum. The period, p, of a pendulum with amplitude α and length L is given by the formula

$$p = 4\sqrt{\frac{L}{g}} K\left(\sin\left(\frac{\alpha}{2}\right)\right),$$

and the classical simple harmonic approximation is that $K(\sin(\alpha/2))$ is close to $\pi/2$ for small angles.

But we digress. Our present interest in elliptic integrals arises through the following function:

$$D(y) = \frac{4yK\left(\sqrt{\frac{(1-3y)(1+y)^3}{(1+3y)(1-y)^3}}\right)}{\sqrt{(1+3y)(1-y)^3}}.$$

This function arises very naturally when physicists study the decay of particles in quantum field theory. Like many of the equations that arise in physics, this looks pretty complicated, and you should be prepared for the fact that the next couple of pages are going to have some even more "symbol-heavy" expressions. Nevertheless, try to hang on; it's a wild ride but a fascinating and rewarding one.

Although the function $D(y)$ leads to many closed-form evaluations, such as

$$\int_0^{1/3} \frac{D(y)}{1-y^2} dy = \frac{\pi^2}{16},$$

in its own right it is somewhat intractable. What happens when we seek help from the computer, asking Maple to supply an ordinary differential equation that D satisfies.

Maple answers this question easily, by using the built-in function `holexprtodiffeq` in the package `gfun`, which takes as input a *holonomic* expression. "A *what* expression?" you cry! Maple's help file tells you they are simple. The hypergeometric representation for D is such an expression, and typing in the Maple instruction `convert (D(y),hypergeom)` will provide it. The function `holexprtodiffeq` now returns a second-order ordinary differential equation, which we shall refer to as DE2.

A reasonable next step is to ask the computer to solve DE2. Working fully formally, Maple's `dsolve` routine says that DE2 is solved by the *HeunG function*. There, Maple has done it again! What on earth is the HeunG function? It turns out that it is a relatively recent, highly-applicable generalization of the hypergeometric function and an even more recent—and very effective—Maple implementation. You may have never heard of it, but Maple knows all about it. Indeed, the Maple help file tells you:

> Heun's equation is an extension of the 2F1 hypergeometric equation in that it is a second-order Fuchsian equation with four regular singular points. The 2F1 equation has three regular singularities. The **HeunG** function, thus, contains as particular cases all the functions of the hypergeometric 2F1 class.
>
> [and much more ...]

There are many solution branches, and with a bit of human intervention, you will eventually discover the identity

$$D(y) = \frac{3\sqrt{3}\pi y}{2} \text{HeunG}(-8, -2; 1, 1, 1, 1; 1 - 9y^2).$$

Let's pause and take stock of where we are. We started with a definition that, although somewhat complicated, at least involved only familiar operations like integrals and square roots. Then we replaced that initial definition with an expression involving a function we'd never heard of until the computer just threw it at us. It looks like we have replaced a hard problem by an even harder one. But remember, the computer is still sitting there, waiting to press ahead. It just needs a little initial direction from an astute human or two, who perhaps recognize a similarity with another situation, and you are away again! In this case, the similarity is with the related function

$$\tilde{D}(y) = \frac{4yK\left(\sqrt{\frac{16y^3}{(1+3y)(1-y)^3}}\right)}{\sqrt{(1+3y)(1-y)^3}}.$$

This has an explicit form in terms of the arithmetic-geometric mean operation (AGM) that always crops up sooner or later when elliptic integrals are concerned. The AGM, as we first saw on page 3, is defined as the common limit of the two sequences (a_n), (b_n), defined by

$$a_0 = a; \quad b_0 = b;$$
$$a_{n+1} = (a_n + b_n)/2; \quad b_{n+1} = \sqrt{a_n b_n}.$$

In terms of the AGM function, we have the elegant, symmetrical, and rapidly computable expression

$$\tilde{D}(y) = 1/\text{AGM}\left(\sqrt{(1-3y)(1+y)^3}, \sqrt{(1+3y)(1-y)^3}\right).$$

When Borwein and Broadhurst first carried out the investigation just described, the thing that led them to shift their attention to this variant of the original D function is that it too can be expressed in terms of that mysterious HeunG function, which in turn led to a series expansion:

$$\tilde{D}(y) = \text{HeunG}(9, 3; 1, 1, 1, 1; 9y^2) = \sum_{k=0}^{\infty} a_k y^{2k}.$$

In this case, Borwein and Broadhurst first identified the coefficients in the power series expansion using Sloane's integer sequence look-up facility, which told them:

$$a_k = \sum_{j=0}^{k} \binom{k}{j}^2 \binom{2j}{j}.$$

The sequence (a_k), which starts 1, 3, 15, 93, 639, ..., is found as sequence A2893 in Sloane's database, and it crops up in a striking number of interesting places. They are called the *hexagonal numbers* because of their role in honeycomb structures.

The $\tilde{D}(y)$ discovery led to the completion of a substantial amount of research, both in quantum field theory and for lattice Green functions in statistical mechanics. A remarkable harvest *within mathematics* was the subsequent evaluation of five rational series in terms of elliptic integrals at so-called *singular values*. Singular values of elliptic integrals are algebraic numbers k_r, such that

$$K\left(\sqrt{1-k_r^2}\right) = \sqrt{r}K(k_r).$$

Here are the five results:

$$\sum_{k=0}^{\infty} \frac{\binom{2k}{k}}{(-108)^k} a_k = \frac{6}{\pi^2}\left(3\sqrt{3} - \sqrt{21}\right) K\left(k_{21}\right) K\left(k_{7/3}\right),$$

$$\sum_{k=0}^{\infty} \frac{\binom{2k}{k}}{(-396)^k} a_k = \frac{6}{\pi^2}\left(3\sqrt{33} - 5\sqrt{11}\right) K\left(k_{33}\right) K\left(k_{11/3}\right),$$

$$\sum_{k=0}^{\infty} \frac{\binom{2k}{k}}{(-2700)^k} a_k = \frac{30}{\pi^2}\left(3\sqrt{57} - 13\sqrt{3}\right) K\left(k_{57}\right) K\left(k_{19/3}\right),$$

$$\sum_{k=0}^{\infty} \frac{\binom{2k}{k}}{(-24300)^k} a_k = \frac{90}{\pi^2}\left(39\sqrt{3} - 7\sqrt{93}\right) K\left(k_{93}\right) K\left(k_{31/3}\right),$$

$$\sum_{k=0}^{\infty} \frac{\binom{2k}{k}}{(-1123596)^k} a_k = \frac{69}{8\pi^2}\left(\sqrt{3} - 1\right)^9 \sqrt{59}K\left(k_{177}\right) K\left(k_{59/3}\right).$$

In fact, there are also alternative forms for all five series in terms of Gamma functions, but in the interest of symbol conservation, we'll omit them here.

And so we have come full circle. Some truly unlikely series evaluations have fallen out of physically driven, computer-assisted analysis. Just imagine if some one asked you to evaluate, say, the last series above and gave you no clues!

Explorations

When a method to compute a mathematical object is implemented in a computer algebra package, it is likely that something more is going on than Spock saying, "Computer, compute to the last digit the value of π." So determining the steps (what educators call "unpacking" the concept) is often a way of learning more.

1. *The Psi function.* Let's look again at the series

$$\zeta(2,1) = \sum_{n=0}^{\infty} \frac{\sum_{k=1}^{n-1} \frac{1}{k}}{n^2},$$

which we introduced in the Explorations for Chapter 4. Typing this into Maple or Mathematica might—depending on the release of the package—return an answer or an unevaluated expression, or it might return

$$\zeta(2,1) = \sum_{n=0}^{\infty} \frac{\Psi(n) + \gamma}{n^2}.$$

If this happens, the package has revealed that it "knows" the *Psi function* (i.e., the logarithmic derivative of the Gamma function, namely $\Gamma'(x)/\Gamma(x)$) and is using a form of telescoping together with the fact that $\Psi(n+1) - \Psi(n) = 1/n$.

This leads to a much more efficient way of computing large sums of harmonic numbers. By summing $\sum_{n=1}^{N} (\Psi(n) + \gamma)/n^2$ numerically, you can estimate $\zeta(3)$ to within roughly $1/N$. This is suggestive for handling numerical computation of more general harmonic sums.

2. *What is κ?* You are presented with

$$\kappa := \sqrt{2} \frac{e^{\sqrt{2}} + 1}{e^{\sqrt{2}} - 1},$$

which begins in decimal as 2.322726139... and has a continued fraction representation [2,3,10,7,18,12,...]. You want to know about κ. What do you do?

Chapter 9

~~→

Take It to the Limit

So put me on a highway
And show me a sign
And take it to the limit one more time
Take it to the limit
Take it to the limit
Take it to the limit one more time
—The Eagles

In the beginning (i.e., freshman mathematics), there is algebra and there is analysis. Analysis, the fledgling mathematician learns, is algebra with limits. Algebra is easy in the early stages; analysis is hard from the get-go—the reason being those limits. As the name suggests, computer algebra systems (CASs), such as Mathematica or Maple, are designed to do algebra. They can also do calculus, because it is mostly a matter of algebraic manipulations according to prescribed rules. But can a CAS do analysis, the subject that lifts the hood on calculus and explains how it works? What that question really comes down to is: how well does a CAS handle limits? More generally, can the methods of experimental mathematics help us with the underlying problems of analysis, namely handling questions about sequences and series? As we shall see, the answer is that CASs and the methods of experimental mathematics can be of considerable assistance when faced with an infinite sequence, an infinite series, or an infinite product.

The default approach to analyze an integer sequence in experimental mathematics is to compute enough initial values for a resource such as Sloane's integer sequence look-up facility to return a plausible looking formula, and then try to prove it is correct, most likely by induction. For example, suppose you are faced with the sequence (u_n), defined by the recursion

$$u_0 = 2,$$

$$u_{n+1} = \frac{2u_n + 1}{u_n + 2},$$

and you want to find a formula for u_n. As a first step, you start to compute. The first few numerators are

2, 5, 14, 41, 122, 365, 1094, 3281, 9842, 29525, 88574

and the denominators are each one less. Sloane's look-up system recognizes the numerator sequence as $(3^n + 1)/2$, in which case,

$$u_n = \frac{3^{n+1} + 1}{3^{n+1} - 1}.$$

Of course, this formula is based on a look-up of a few numerical values. But now that you have the formula, it's an easy matter to prove it by induction.

When it comes to computing the sum of an infinite series, the default experimental mathematics approach is to compute a sufficiently precise numerical sum and then try to recognize the result using a tool such as the ISC or the PSLQ algorithm. For instance, using this approach, Gregory and David Chudnovsky provided each of the following series evaluations:

$$\sum_{n=0}^{\infty} \frac{50n - 6}{2^n \binom{3n}{n}} = \pi,$$

$$\sum_{n=0}^{\infty} \frac{2^{n+1}}{\binom{2n}{n}} = \pi + 4,$$

$$\sum_{n=0}^{\infty} \frac{(4n)!(1 + 8n)}{4^{4n} n!^4} = \frac{2}{\pi\sqrt{3}},$$

$$\sum_{n=0}^{\infty} \frac{\binom{2n}{n}}{n^2 4^n} = \frac{\pi^2}{6} - 2\log^2 2$$

(and a whole lot more in a similar vein).

On occasion, a CAS is simply "too smart," giving a result that is of little help to the human user, but a little ingenuity can sometimes save the day.

For instance, suppose you are faced with evaluating the infinite product

$$\prod_{n=2}^{\infty} \frac{n^3 - 1}{n^3 + 1}.$$

Mathematica returns an expression involving the Gamma function, while Maple gives the answer 2/3. Whereas Maple's answer is undoubtedly the simpler—and how!—in neither case are you any the wiser as to what is going on. To gain understanding, you can try evaluating the finite product and then taking the limit. If you do that using Maple, you will get an answer involving the Gamma function that can be simplified to

$$\prod_{n=2}^{N} \frac{n^3 - 1}{n^3 + 1} = \frac{2}{3} \frac{N^2 + N + 1}{N(N + 1)}.$$

Staring at this answer for a while, you eventually hit upon the idea of a "telescoping" argument (successive terms that cancel), and you quickly arrive at the following derivation:

$$\prod_{n=2}^{N} \frac{n^3 - 1}{n^3 + 1} = \prod_{n=2}^{N} \frac{(n - 1)(n^2 + n + 1)}{(n + 1)(n^2 - n + 1)}$$

$$= \frac{\prod_{n=0}^{N-2} (n + 1)}{\prod_{n=2}^{N} (n + 1)} \cdot \frac{\prod_{n=2}^{N} (n^2 + n + 1)}{\prod_{n=1}^{N-1} (n^2 + n + 1)}$$

$$= \frac{2}{N(N + 1)} \cdot \frac{N^2 + N + 1}{3} \to \frac{2}{3}.$$

Ah, so that's what is going on! It's of interest to note that the seemingly simpler product with square powers instead of cubes evaluates to a transcendental value:

$$\prod_{n=2}^{\infty} \frac{n^2 - 1}{n^2 + 1} = \frac{\pi}{\sinh \pi}.$$

(Maple gives this answer directly; Mathematica again gives a result involving the Gamma function.) It's a nice challenge to determine what happens with fourth powers.

It sometimes helps to use the computer to draw a graph. For example, the following problem appeared in the *American Mathematical Monthly*, Vol. 108, in 2001. Define a sequence (a_n) of fractions by setting $a_1 = 1$, and producing a_{n+1} by replacing each fraction $1/d$ in the expression for a_n by $1/(d+1) + 1/(d^2 + d + 1)$. The sequence thus starts like this:

$$a_2 = \frac{1}{2} + \frac{1}{3}, \quad a_3 = \frac{1}{3} + \frac{1}{7} + \frac{1}{4} + \frac{1}{13},$$

$$a_4 = \frac{1}{4} + \frac{1}{13} + \frac{1}{8} + \frac{1}{57} + \frac{1}{5} + \frac{1}{21} + \frac{1}{14} + \frac{1}{183}.$$

The question is, what is the limit of this sequence?

The trick is to look at the functions $s_n(x)$ defined on positive real numbers x by:[1]

$$s_0(x) = \frac{1}{x}; \quad s_{n+1}(x) = s_n(x+1) + s_n(x^2 + x + 1).$$

Clearly, $a_n = s_{n+1}(1)$. If we graph the $s_n(x)$, they look like reciprocal functions. Graphing the functions $s_n(1/x)$ instead displays a family that appears to converge rapidly to a smooth, monotone increasing function we'll call $g(x)$, shown in Figure 4. (We can check the rate of convergence by comparing, say, $s_{24}(1/x)$ with $s_{25}(1/x)$, which agree to around four decimal places.) What is the function $g(x)$?

Looking at the sequence of calculated numerical values used for plotting, we find that, while $g(x) = \lim_{n \to \infty} s_n(1/x)$ is not defined at zero, it appears that $\lim_{x \to 0} g(x) = 0$. See Figure 4.

It also looks as if $g'(0) = 1, g(1) \approx 0.7854$, and $g'(1) = 1/2$. The value 0.7854 looks suggestively like an approximation to $\pi/4$. Perhaps $g(x) = \arctan(x)$?

Well, it's definitely worth a try. Let $f(x) = \arctan(1/x)$ for $x > 0$. By applying the addition formula for the tangent, we get

$$\tan\left[f(x+1) + f(x^2 + x + 1)\right] = \frac{\dfrac{1}{x+1} + \dfrac{1}{x^2 + x + 1}}{1 - \dfrac{1}{x+1}\dfrac{1}{x^2 + x + 1}} = \frac{1}{x} = \tan\left[f(x)\right].$$

---

[1]This is much easier when working in a computer algebra system than elsewhere. We can think of the function instead of the number, and at least for a while the system can happily compute the explicit functional iteration.

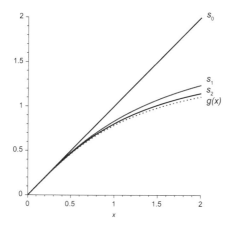

Figure 4. The crucial graph that leads to the solution of the *Monthly* problem.

Thus $f(x)$ satisfies $f(x) = f(x+1) + f(x^2 + x + 1)$. Aha! All we have to do now to finish is show that $s_n(x)$ converges pointwise to $f(x)$.

The first step is to verify that the function $E(x) = 1/(x f(x))$ decreases strictly to 1 as x goes to infinity. By differentiation, it suffices to show that $-\arctan(x) + x/(x^2 + 1) < 0$. But this follows from the fact that $-\arctan(x) + x/(x^2 + 1)$ is strictly decreasing (its derivative is $-2x^2/(x^2 + 1)^2$) and starts at 0 for $x = 0$.

The next step is to show that for all $x > 0$,

$$f(x) \le s_n(x) \le f(x)E(x+n). \tag{9.1}$$

We do this by induction. For $n = 0$, this is merely the valid inequality $x f(x) \le 1$. Assuming the inequality for some $n > 0$, we infer

$$f(x+1) \le s_n(x+1) \le f(x+1)E(x+n+1)$$

and, using the monotonicity of E,

$$f(x^2 + x + 1) \le s_n(x^2 + x + 1) \le f(x^2 + x + 1)E(x^2 + x + 1 + n)$$
$$\le f(x^2 + x + 1)E(x + n + 1).$$

Adding and using the functional equation for f, we get (9.1) for $n + 1$.

That completes the proof, and now we know that $a_n \to \pi/4$. Definitely worth a pat on the back!

Finally, we'll take a look at work done by Borwein and Roland Girgensohn on the family of infinite binomial sums of the form

$$b(k) = \sum_{n=1}^{\infty} \frac{n^k}{\binom{2n}{n}}$$

for nonnegative integers k. They used experimental techniques to derive closed forms for these sums.

The key observation is that the sums have integral representations involving the polylogarithms

$$Li_p(z) = \sum_{n=1}^{\infty} \frac{z^n}{n^p}$$

that we met once already in Chapter 5.

We start with something called the Beta function, which is defined by

$$\beta(x,y) = \int_0^1 t^{x-1}(1-t)^{y-1} dt.$$

For $x, y > 0$, this function can be expressed neatly as

$$\beta(x,y) = \frac{\Gamma(x)\Gamma(y)}{\Gamma(x+y)}.$$

(A nice consequence is that $\Gamma(1/2) = \sqrt{\pi}$.) This representation is very useful for expressing reciprocals of binomial coefficients. In particular, writing

$$\frac{1}{\binom{2n}{n}} = (2n+1)\beta(n+1,n+1) = n\beta(n,n+1),$$

we discover that

$$b(k) = \int_0^1 Li_{-k}(x(1-x)) dx = 2Li_{-k-1}(x(1-x))$$

$$= \int_0^1 \frac{Li_{-k-1}(x(1-x))}{x} dx. \qquad (9.2)$$

For a given k, this integral is easy to compute symbolically in Maple, and, with some additional effort, in Mathematica.

$Li_{-k}(x)$ is clearly a rational function, and thus it can be written as a partial fraction

$$Li_{-k}(x) = \sum_{j=1}^{k+1} \frac{c_j^k}{(x-1)^j}.$$

Since $xdLi_{-k}(x)/dx = Li_{-k-1}(x)$, we obtain the recursion

$$c_j^k = -(jc_j^{k-1} + (j-1)c_{j-1}^{k-1}). \tag{9.3}$$

Letting

$$M(k,x) = Li_{-k}(x) + 2Li_{-k-1}(x),$$

we may easily verify that the coefficients of the partial fraction of M are governed by the recursion (9.3), with initial conditions given by $c_1^0 = 1, c_2^0 = 2$, otherwise $c_j^0 = 0$. We may then use a CAS to show that

$$c_j^k = \frac{(-1)^{k+j}}{j} \sum_{m=1}^{j} (-1)^m (2m-1) m^{k+1} \binom{j}{m}.$$

(The hard part is coming up with this in the first place; the verification is much less of a challenge, and could be done by hand.)

Noting that the value of the integral in (9.2) is of the form

$$b(k) = \sum_j c_j^k \int_0^1 (1 - x(1-x))^{-j} dx,$$

we look for a recursion for

$$d(j) = \int_0^1 (1 - x(1-x))^{-j} dx.$$

By computing the first few cases, we determine that $d(j)$ is a rational combination of 1 and $\pi/\sqrt{3}$. Thus, it is reasonable to hunt for a two-term recursion for d. We use an integer relation algorithm to find a linear relation between $d(p)$, $d(p+1)$, and $d(p+2)$, for $0 \le p \le 4$. When we do, we obtain the relations

$$[2,2,-3], [-2,9,-6], [6,-16,9], [-10,23,-12], [14,-30,15].$$

By inspection, $d(0) = 1$, $d(1) = 2\pi/3\sqrt{3}$, and for $p \geq 2$,

$$(4p - 10)d(p - 2) - (7p - 12)d(p - 1) + (3p - 3)d(p) = 0. \qquad (9.4)$$

Having discovered the recursion, we can prove it by looking at the indefinite integral from 0 to t, which both Mathematica and Maple can perform happily. We can then verify that the integral has a zero at $t = 1$. Specifically, combining the integrals in (9.4) gives

$$\int_0^t \left(\frac{4p - 10}{(1 - x(1 - x))^{p-2}} - \frac{7p - 12}{(1 - x(1 - x))^{p-1}} + \frac{3p - 3}{(1 - x(1 - x))^p} \right) dx$$
$$= -\frac{(2t - 1)(t - 1)t}{(1 - t + t^2)^{p-1}}$$

$$(9.5)$$

and then differentiating this last expression and simplifying yields the integrand as required. Since the right-hand side of (9.5) has a zero at $t = 1$, we are done. (In fact, the quantities $d(j)$ can be computed explicitly.)

This shows that

$$b(k) = p_k + q_k \frac{\pi}{\sqrt{3}}$$

with easily and prescriptively computable rationals p_k, q_k. In particular, the first three sums are

$$\sum_{n=1}^{\infty} 1 \Big/ \binom{2n}{n} = \frac{1}{3} + \frac{2}{9} \frac{\pi}{\sqrt{3}},$$

$$\sum_{n=1}^{\infty} n \Big/ \binom{2n}{n} = \frac{2}{3} + \frac{2}{9} \frac{\pi}{\sqrt{3}},$$

$$\sum_{n=1}^{\infty} n^2 \Big/ \binom{2n}{n} = \frac{4}{3} + \frac{10}{27} \frac{\pi}{\sqrt{3}}.$$

Cool, *n'est-ce pas?*

Take it to the limit, Take it to the limit, Take it to the limit one more time ...

Uneasy riders crossing the limit

Explorations

1. *Finding limits.*

(a) Let $a_0 = 0$, $a_1 = 1/2$ and define

$$a_{n+1} = \frac{\left(1 + a_n + a_{n-1}^3\right)}{3}$$

for $n > 1$. Determine the limit and find out what happens when $a_1 = a$ is allowed to vary.

(b) Let $a_1 = 1$ and define

$$a_{n+1} = \frac{3 + 2a_n}{3 + a_n}.$$

Again determine the limit and find out what happens when $a_1 = a$ is allowed to vary.

These two limits are easy enough to find and (depending on what you know) to prove.

(c) Let $a_1 \geq 1$ be given and determine the limit of the iteration

$$a_{n+1} := a_n - \frac{a_n}{\sqrt{1 + a_n^2}} + \sin(\theta)$$

for arbitrary θ.

2. *Hirschhorn's limit.* The next limit, which was studied by Mike Hirschhorn, should be easy to find but is hard to prove. Identify the limit of M_n as n tends to infinity, where

$$M_n := n \frac{\int_0^1 t^{n-1} (1 + t)^n dt}{2^n}.$$

3. *Mean iterations.* By contrast, the following limits are harder to find, but easier to prove once discovered. A *(strict) mean* $M(a, b)$ is a continuous function of two positive numbers that outputs a number c lying between a and b (strictly so, if a is not equal to b). (The arithmetic and geometric means are clearly such objects.) A mean iteration takes two means, M and N, and iterates by setting $a_0 := a, b_0 := b$, and

$$a_{n+1} := M(a_n, b_n), \quad b_{n+1} := N(a_n, b_n).$$

The limit of such a strict mean iteration always exists and is denoted by $M \otimes N(a, b)$. Thus, the AGM of Gauss that we explored before can be written as $A \otimes G$. Because both A and G are symmetric, convergence is fast. In the following two cases, the challenge is to identify the limit.

(a) $M(a, b) := (a + \sqrt{ab})/2, \quad N(a, b) := (b + \sqrt{ab})/2,$

(b) $M(a, b) := (a + b)/2, \quad N(a, b) := 2ab/(a + b),$

so that N is the *harmonic mean*, usually written H. In the latter case convergence is quadratic, in the former it is linear.

4. *Failure of L'Hopital's rule.* Consider the functions defined by

$$f(x) := x + \cos(x)\sin(x), \quad g(x) = e^{\sin(x)}(x + \cos(x)\sin(x)).$$

Confirm that, despite the fact that

$$\lim_{x \to \infty} \frac{f'(x)}{g'(x)} = 0,$$

the corresponding limit $\lim_{x \to \infty} f(x)/g(x)$ does not exist. Humans and machines are both prone to carelessly divide by zero and the like. Such is the price of working heuristically.

Chapter 10

⤳

Danger! Always Exercise Caution When Using the Computer

> *Computers are useless. They can only give you answers.*
> —Pablo Picasso (1881–1973)

It's a nice quote, but Picasso clearly didn't understand the potential of computers when he made that remark. The pioneering computer scientist R. W. Hamming was closer to the mark when he observed: "The purpose of computing is insight, not numbers." Still, like any powerful tool, the computer can be dangerous if not used with care. This chapter is the obligatory customer protection warning.

We've all heard airline stewards say it so many times that it's practically become a cliché: Exercise caution when opening the overhead bins. The same advice holds when carrying out an investigation using techniques of experimental mathematics. Failure to heed this warning can lead to you being hit over the head by a surprise case—on the airplane it's a suitcase, in mathematics it's a misleading case. On both occasions the impact can be painful.

For example, if you have access to a computer algebra system or a multi-precision scientific calculator program (several of which you can download for free from the Web), and you compute

$$e^{\pi\sqrt{163}}$$

to 30 places of accuracy, you will obtain the result

262537412640768744.000000000000,

at which point you may become very excited. Aware of Euler's famous identity

$$e^{i\pi} = -1,$$

a remarkable result that relates the four fundamental mathematical constants e, π, i, and 1, the first two of which are known to be irrational, you see future fame as the person who has discovered another amazing relation between e and π.

But then the doubts set in. First, you remember that Euler's identity is a bit misleading—though no less remarkable for that—since there are actually infinitely many different solutions to the equation $e^{ix} + 1 = 0$. Then there is that integer 262537412640768744. Somehow, it seems too arbitrary, too "unspecial," to be the value of $e^{\pi\sqrt{163}}$. In fact, it is not. Twelve decimal places all equal to zero is highly suggestive, but if you increase the precision to 33 places, you find that

$$e^{\pi\sqrt{163}} = 262537412640768743.999999999999250\ldots$$

This is still an interesting occurrence, and you would be right to suspect that there is something special about that number 163 that makes this particular power of e so close to an integer. We won't go into that here—it's an algebra thing, outside the scope of this book. Our point is simply to illustrate that computation can sometimes lead to conclusions that turn out to be incorrect. In this case the error lies in mistaking for an integer a number that is in fact transcendental.

Another coincidence, in this case most likely having no mathematical explanation, is that to ten significant figures

$$e^{\pi} - \pi = 19.99909997 \approx 20.$$

Here is another example where the numbers can be misleading. If you use a computer algebra system to evaluate the infinite sum

$$81 \sum_{n=1}^{\infty} \frac{\lfloor n \tanh \pi \rfloor}{10^n},$$

you will get the answer 1. But this answer is incorrect. The series actually converges to a transcendental number, which to 268 decimal places is equal to 1. In other words, you would need to calculate the numerical value to at least 269 places to determine that it is not 1, although to ensure there were no rounding errors, you would want to carry out the calculation to an even greater accuracy. Behind this error is the fact that

$0.99 < \tanh \pi < 1$, and hence $\lfloor n \tanh \pi \rfloor$ will be equal to $n - 1$ for many n; in fact, for $n = 1, \ldots, 268$.

This example highlights yet again the danger that is ever-present in experimental mathematics: getting an answer that looks like something it is not.

Or try this one on for size (and what a size it is in terms of accuracy). The following "equality" is correct to *over half a billion* digits:

$$\sum_{n=1}^{\infty} \frac{\left\lfloor ne^{\pi\sqrt{163}/3} \right\rfloor}{2^n} = 1,280,640.$$

But this sum, far from being an integer (*conceptually* far, that is!), is provably irrational, indeed transcendental. As you have probably guessed, this is a "cooked" example, related to our first example. But what an example it is!

Here is another cautionary tale. Euler defined the *trinomial numbers* by

$$t_n = \left[x^0\right] \left(x + 1 + \frac{1}{x}\right)^n,$$

where $[x^k] P(x)$ denotes the coefficient of x^k in a polynomial or power series $P(x)$.

There are alternative definitions of t_n, as a closed form, via a generating function, and through a three-term recursion.

Euler observed that for $n = 0, 1, \ldots, 7$,

$$3t_{n+1} - t_{n+2} = F_n(F_n + 1),$$

where F_n is the nth Fibonacci number. Given the existence of a recursive definition, it would be tempting to guess that this equality holds in general, but that is not the case.

Another example where a pattern holds for the first few cases but then fails involves that old favorite example from your first course on real analysis, the function

$$\operatorname{sinc}(x) = \sin x / x.$$

Robert Baillie discovered recently that the identity

$$\sum_{n=1}^{\infty} \text{sinc}^N(n) = -\frac{1}{2} + \int_0^{\infty} \text{sinc}^N(x)dx$$

holds for $N = 1, 2, 3, 4, 5,$ and $6,$ but fails at 7.

Still, few experienced mathematicians would be persuaded by a mere seven or eight instances, even if they find those results suggestive. Absent any convincing circumstantial evidence, they would want more cases than that. But our confidence tends to get stronger when the number of cases gets up into the thousands. Nevertheless, we still need to exercise caution.

For example, consider the sequence u_n defined by

$$u_{n+2} = \left\lfloor 1 + u_{n+1}^2 \Big/ u_n \right\rfloor,$$

$$u_0 = 8, \quad u_1 = 55.$$

and the rational function

$$R(x) = \frac{8 + 7x - 7x^2 - 7x^3}{1 - 6x - 7x^2 + 5x^3 + 6x^4}$$

$$= 8 + 55x + 379x^2 + 2612x^3 + \dots$$

(Don't ask! You tend to keep coming across this kind of example when you do experimental mathematics.)

If you compute some values (something your grandparents would have done with pencil and paper on a rainy afternoon, but is now done on machines costing thousands of dollars that do it while playing music for us at the same time), you will find that

$$u_n = [x^n] R(x)$$

for all n up to 10,000—at which point you might be tempted to conjecture that the result is universally valid. (Your grandparents would have reached that conclusion well before then.) But that's not the case. The equation fails for the first time at $n = 11,056$. (Remember that number in case you ever find yourself ill in bed and a colleague comes to visit you and, by way of trying to take your mind off your condition, says, "I came here in a taxi with the license number 11056. That seemed to me a completely uninteresting number." Just imagine how impressive your reply will seem.)

Okay, we're on a roll now. Try this one on for size. Imagine you are given a homework assignment to evaluate the integrals:[1]

$$I_1 = \int_0^\infty \text{sinc}(x)dx,$$

$$I_2 = \int_0^\infty \text{sinc}(x)\text{sinc}\left(\frac{x}{3}\right) dx,$$

$$I_3 = \int_0^\infty \text{sinc}(x)\text{sinc}\left(\frac{x}{3}\right)\text{sinc}\left(\frac{x}{5}\right) dx,$$

$$\dots$$

$$I_8 = \int_0^\infty \text{sinc}(x)\text{sinc}\left(\frac{x}{3}\right)\text{sinc}\left(\frac{x}{5}\right)\dots\text{sinc}\left(\frac{x}{15}\right) dx.$$

You set your favorite computer algebra system to work and you find that $I_1 = \dots = I_7 = \pi/2$. Okay, you get the picture, why spend time on the last one? Well, you decide to make sure. Here goes.

"What?!" Your screen has just filled up with the following unexpected output:

$$I_8 = \frac{467807924713440738696537864469}{935615849440640907310521750000}\pi$$

$$\approx 0.499999999992646\pi.$$

Clearly, you have made a keyboard error when you entered the final problem. So you try again, checking carefully to make sure there are no typos. You get exactly the same output, at which point, you contact the developer of the computer algebra system and say you have found an obscure arithmetic bug. This is exactly what the researcher who first found the above result did, and the software developer agreed that this was clearly a software bug.

The suspicion of a bug, if anything, was bolstered by that fact that the integral I_9 is beyond the capabilities of the CAS used, which did not return an answer.

---

[1] Remember from a couple of pages ago that $\text{sinc}(x) = \sin x/x$.

But in fact there is no bug. The value the CAS returned for I_8 is correct. What was faulty was your conclusion that (1) the pattern established by the first seven integrals would continue, and (2) that kind of integral never gives an answer like 0.499999999992646π.

It's all in a day's work for the experimental mathematician.

Here's another misleading calculation, this time involving the function

$$C(x) = \cos(2x) \prod_{n=1}^{\infty} \cos\left(\frac{x}{n}\right).$$

If you set your computer to evaluate the integral

$$I = \int_0^{\infty} C(x)dx$$

numerically and you work hard and cleverly, you will get an answer that agrees with $\pi/8$ to 40 decimal places. But a careful hybrid of numeric plus symbolic integration allows you to estimate the error and show that $I < \pi/8$.

Here is another instance where it pays to be cautious. Given a natural number n, let $e(n)$ be the number of even decimal digits of n and $o(n)$ the number of odd digits. A somewhat intricate calculation, which we'll give in a moment, shows that

$$\sum_{n=1}^{\infty} \frac{o(2^n)}{2^n} = \frac{1}{9}.$$

Okay, now that you know that, what would you think is the value of the following sum?

$$\sum_{n=1}^{\infty} \frac{e(2^n)}{2^n}$$

Hands up all those who said 1/9, or gave some other small fraction. If you didn't raise your hand, it was almost certainly because this example is in the context of this chapter. In fact, the correct answer is

$$\sum_{n=1}^{\infty} \frac{e(2^n)}{2^n} = \sum_{n=1}^{\infty} \frac{\lfloor n\log_{10} 2\rfloor +1}{2^n} - \sum_{n=1}^{\infty} \frac{o(2^n)}{2^n},$$

which is transcendental. (A quadruple-precision computation will lead you to suspect the answer is 3166/3069.) We won't prove the transcendency here, but here is a proof of the $o(2^n)$ result.

Let $0 < q < 1$ and take $m \in \mathbb{N}$, $m > 1$. Consider the base-m expansion of q:

$$q = \sum_{n=1}^{\infty} \frac{a_n}{m^n}, \text{ with } 0 \le a_n < m,$$

where, when ambiguous, we take the terminating expansion. Then a_n is the remainder of $\lfloor m^n q \rfloor$ modulo m, and so we may write

$$q = \sum_{n=1}^{\infty} \frac{\lfloor m^n q \rfloor \pmod{m}}{m^n}.$$

Now let

$$F(q) = \sum_{k=1}^{\infty} c_k q^k = \sum_{k=1}^{\infty} c_k \sum_{n=1}^{\infty} \frac{\lfloor m^n q^k \rfloor \pmod{m}}{m^n} = \sum_{n=1}^{\infty} \frac{f(n)}{m^n}$$

with

$$f(n) = \sum_{k \ge 1} c_k \left(\lfloor m^n q^k \rfloor \pmod{m} \right).$$

If $q = 1/b$, where b is an integer multiple of m, then $\lfloor m^n q^k \rfloor \pmod{m}$ is the kth digit mod m of the base-b expansion of the integer m^n. (We start the numbering of the digits at 0. For example, the zeroth digit of 1205 is 5.) Thus, for $F(q) = q/(1 - q)$ and $m = 2$ (and b even), $f(n)$ counts the odd digits in the base-b expansion of 2^n. For $b = 10$, we have $f(n) = o(2^n)$, giving

$$\frac{1}{9} = F\left(\frac{1}{10} \right) = \sum_{n=1}^{\infty} \frac{o(2^n)}{2^n},$$

as required.

Finally, we recall the famous *Skewes number*, related to (early work on) the Prime Number Theorem, which provides an upper bound for the first value of x for which the inequality

$$\int_2^x \frac{dt}{\log t} \ge \pi(x)$$

fails, where $\pi(x)$ is the number of primes below x. In 1933, Stanley Skewes showed that $10^{10^{10^{34}}}$ is an upper bound. The best current result is that the first known cross-over occurs around $1.397162914 \times 10^{316}$. While considerably smaller than Skewes's bound, this number still far exceeds anything with astronomical significance.

This classic result is an excellent reminder of the advice we opened the chapter with: in experimental mathematics, always exercise caution!

Explorations

Computer algebra systems are often criticized for doing one of two things: over simplifying or under simplifying expressions. Getting the balance right is not easy and is a matter of balancing user expectations with what can be proved. A user probably does not want to see $\sqrt{\cos(\theta)^2}$ as an answer, but it may not be clear to the system where θ lies, even if it is clear to the user. You probably often want

$$\sqrt{-x} = i\sqrt{x},$$

but you surely do not want to obtain

$$3 = \sqrt{-(-9)} = i\sqrt{-9} = i \cdot i\sqrt{9} = -3.$$

Often a CAS will have used various functional equations and transformations to compute an object, and so it may return a nice value for a divergent series. Different systems have different conventions about the domains of inverse trig functions. Keep your eyes open.

1. *Simplification.* Simplify the following two radicals:

   (a) $\alpha_1 := \sqrt[3]{\cos(2\pi/9)} + \sqrt[3]{\cos(4\pi/9)} + \sqrt[3]{\cos(8\pi/9)}$
   (b) $\alpha_2 := \sqrt[3]{\cos(2\pi/7)} + \sqrt[3]{\cos(4\pi/7)} + \sqrt[3]{\cos(6\pi/7)}$

   Hint: Try to identify them from their numerical values.

2. *Recent American Mathematical Monthly problem.* A recent *Monthly* problem[2] is equivalent to evaluating

$$\sigma(m,n) := \sum_{k=0}^{m} 2^k \binom{2m-k}{m+n} + \sum_{k=0}^{m} \binom{2m+1}{m+k}$$

---

[2]*American Mathematical Monthly*, February 2007, #11274.

for nonnegative integers m and n. This is mathematically the same as

$$\sigma^*(m,n) := \sum_{k=0}^{\infty} 2^k \binom{2m-k}{m+n} + \sum_{k=0}^{\infty} \binom{2m+1}{m+k},$$

but a CAS may not think so. What is the correct answer?

Chapter 11

~~→

Stuff We Left Out (Until Now)

Please, sir, I want some more.
—Oliver Twist, in the novel by Charles
Dickens (1786–1851)

As we said in Chapter 1, this book is not meant to provide comprehensive coverage of experimental mathematics. Indeed, it's not clear what such a book would look like—except to say that it would be very big—since experimental mathematics is really an *approach* to mathematical discovery. (That approach does, however, imply a view on what constitutes mathematical *knowledge*, a view that goes well beyond the traditional "what has been proved" to encompass as well "that for which we have good evidence"—with the same caveat on the latter as is widely accepted in the natural sciences.)

We have kept our focus very much on the use of experimental methods in real analysis, analytic number theory, and calculus, using discoveries in those areas to illustrate and exemplify the experimental approach. In this final chapter, we try to redress the balance a bit by widening the scope and looking at other parts of the subject.

A Picture May Be Worth a Thousand Symbols

Suppose that you needed to know which of the two functions $y - y^2$ and $-y^2 \log y$ is larger on the unit interval. What about the pair $y^2 - y^4$ and $-y^2 \log y$? You could use traditional analytic methods—and if you wanted a rigorous *proof*, that would be the way to go. But if you just wanted to know the answer, the fastest way is to use a computer or a graphing calculator to draw the curves. When you do this, you get the two displays shown in Figure 5, and the question is answered.

Figure 5. Graphical comparison of $y - y^2$ to $-y^2 \log y$ (left) and $y^2 - y^4$ to $-y^2 \log y$ (right).

Discovery by Visualization

Sometimes, drawing a picture—more precisely, finding a way to draw "the right" picture—yields more than the solution to a technical problem like the one we just looked at; it provides the key insight to a major discovery. A spectacular early example where computer graphics led to a deep discovery occurred in 1983, when mathematicians David Hoffman and William Meeks III discovered a new minimal surface.

A minimal surface is the mathematical equivalent of an infinite soap film. Real soap films stretched across a frame always form a surface that occupies the minimal possible area. The mathematical analog is a minimal surface that stretches out to infinity. Such surfaces have been studied for over two hundred years, but until Hoffman and Meeks made their discovery, only three such surfaces were known. Today, as a result of using visualization techniques, mathematicians have discovered many such surfaces. Much of what is known about minimal surfaces is established by more traditional mathematical techniques. But, as Hoffman and Meeks showed, the computer graphics can provide the mathematician with the initial discovery as well as the intuition needed to find the right combination of those traditional techniques.

"We were surprised that computer graphics could actually be used as an exploratory tool to help us solve the problem," says Hoffman. "The surface couldn't be understood until we could see it. Once we saw it on the screen, we could go back to the proof."

The type of surface that Hoffman and Meeks investigated is a "complete, embedded, minimal surface of finite topology." The term "complete" here means that the surface, roughly speaking, has no boundaries. A smooth plane that extends in all directions without end is one example of a complete surface. It also happens to be a minimal surface because putting any kind of fold into the plane increases its surface area. Another example of a complete minimal surface is the catenoid, which looks like an infinitely extended hourglass. (The soap film that connects two parallel circles of wire as they are pulled apart looks like the central piece of a catenoid.) Both the plane and the catenoid are also "embedded" surfaces—they do not fold back and intersect themselves. Another surface with these properties is the helicoid—imagine a soap film stretching along the curves of an infinitely long helix or spiral.

Until the work of Hoffman and Meeks, these three were the only known examples of complete, embedded, minimal surfaces (of finite topology). A few mathematicians had speculated that these were the only possible examples.

But then Hoffman started to look at the equations for a surface first written down by a Brazilian graduate student, Celso J. Costa, in his doctoral thesis. Costa was able to prove that this particular surface is minimal and complete. Hoffman suspected that it could be embedded. Mathematical clues suggested that the surface contained two catenoids and a plane that all somehow sprouted from the center of the figure. But that was not enough to show what it looked like.

Enter the computer, used both to compute numerical values for the surface's coordinates and to draw pictures of its core. The big question was whether Costa's surface intersected itself. If it did, then the surface would not be embedded and that would be the end of the matter. If there was no visible evidence of an intersection, however, then he could go ahead with trying to prove that the surface really was embedded.

The first pictures did indicate that the surface was free of self-intersections. Seeing the surface from different points of view also showed that it had a high degree of symmetry, but it took "extended staring" over several days to piece together the true form of this new minimal surface, says Hoffman. "How it fitted together was not obvious." Until, that is, the human-machine interaction produced the image shown in Figure 6. (Figure 6 is taken from the peer-reviewed Electronic Geometric Models site (http://www.eg-models.de/), which is an exemplar of manipulable visual resources to come.)

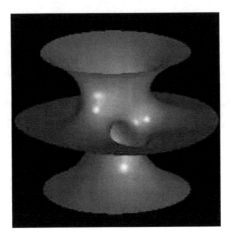

Figure 6. Computer-generated image of the Costa-Hoffman-Meeks surface.

Whereas Hoffman and Meeks's work comprised the experimental use of a computer to understand a mathematical object sufficiently well to find a traditional proof, another application of experimental techniques went in the opposite direction: computer graphics, used experiment-ally, resulted in a deeper understanding of a result that had already been proved using traditional techniques. The topic was knot theory.

A Knotty Problem

For a general background in knot theory, we refer to Colin Adams's *The Knot Book* [Adams 94].

In early knot tables, the two knots shown in Figure 7 were listed as separate knots. (Some knot tables still list them as different.)

In his book *Knots and Links* [Rolfsen 76], where the two knots are listed as 10_{161} and 10_{162}, author Dale Rolfsen notes, however, that in 1974, Kenneth Perko proved that these two knots are the same [Perko 74]. A natural question is, what sequence of basic moves (specifically, the famous Reidemeister moves, which are used to rearrange knots) will transform one of these knots into the other? Both in earlier times and today, knot theorists sometimes construct physical knots (rubber tubing is a popular material

Figure 7. For many years, these figures were thought to depict different knots, but in fact they are equivalent.

Figure 8. Each knot diagram shown in Figure 7 can be deformed into this knot.

that lends itself well to this process) to seek insight into knot equivalences. Recently, computer graphics packages have been added to the knot theorist's arsenal. One such is KnotPlot, available at http://knotplot.com. If you go to http://knotplot.com/perko, you will find a lengthy sequence of images showing the equivalence of 10_{161} and 10_{162}. The sequence was discovered experimentally, with the deformations now being performed entirely automatically using the KnotPlot tool. In fact, both may be deformed to the knot shown in Figure 8.

The Proof of the Four Color Theorem

In the work by Hoffman and Meeks on minimal surfaces, which we looked at a moment ago, the initial groundwork involved experimental investiga-

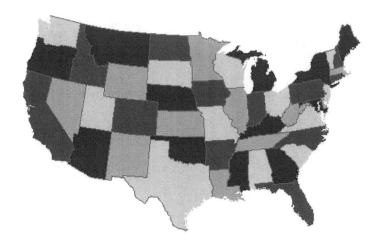

Figure 9. Map of the US colored using four colors.

tion using a computer, but the final result was a traditional mathematical proof. Things turned out differently with another major breakthrough in mathematics. The proof of the Four Color Theorem not only involved experimental work using a computer, but an essential part of the final proof was of necessity carried out by computer.

First formulated in 1852, the four color conjecture asserted that, to color *any* map (drawn on a plane), subject to the obviously desirable requirement that no two regions (countries, counties, etc.) sharing a length of common boundary should be given the same color, the maximum number of colors required is four. For example, the US map in Figure 9 uses only four colors (or four shades of gray).

It is easy to see that some maps require at least four colors, such as the map shown in Figure 10.

It did not take long for various people to discover that five colors are enough for any map. But a century of attempts to show that four is always adequate failed. The history of failed attempts, coupled with the problem being so easy to understand, led to it becoming probably the second most famous unsolved puzzle in mathematics after Fermat's last theorem.

Then, in 1976, two mathematicians at the University of Illinois, Kenneth Appel and Wolfgang Haken, announced that they had proved it.

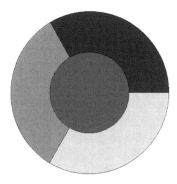

Figure 10. A simple map that cannot be colored using three colors.

That alone was a major news story. But for mathematicians, the really surprising aspect was that crucial parts of Appel and Haken's argument were carried out by a computer, using ideas that had themselves been formulated as a result of computer-based evidence. For the first time in the history of mathematics, what constitutes a mathematical proof had suddenly changed. Before 1976, a *proof* had been a logically sound piece of reasoning by which one mathematician could convince another of the truth of some assertion. By reading a proof, a mathematician could become convinced of the truth of the statement in question, and also come to understand the reasons for its truth. Of course, with only so much time available, individual mathematicians often left it to others whom they trusted (such as journal referees) to check the details. But proofs were inherently generated and verified by human brains.

This was not the case with the proof of the Four Color Theorem. The use of the computer was absolutely essential. To accept the proof, you had to believe that the computer program did what its authors claimed.[1]

There were actually two ways Appel and Haken made use of the computer. One way was to check a large number of cases. Each case on its own could be checked by a human, but there were too many for one per-

---

[1] As it turned out, researchers subsequently found flaws in the original proof, and at least one later computer-based proof was carried out to deal with the short-comings of the first. Any remaining uncertainty was finally dispelled in 1997, when Robertson, Seymour, and Thomas produced a computer-assisted proof that was computer verifiable as correct.

son to check (and remain sane). The other was to generate those cases in the first place. This latter part of the process was one of the first examples of experimental mathematics (in the sense that we have been using the term).

The basic idea for the Appel-Haken proof emerged during the early work on the problem in the nineteenth century. You start by assuming there is a map for which five colors are required (i.e., four colors are definitely not sufficient). There will then be one such map with the minimal number of regions. With a bit of reflection, you can show that there will be such a minimal counterexample (to the conjecture) map having certain nice properties. Then you find a region that can be removed from the map without changing the number of colors required to color it. If you can show that there is always such a region, you are done by contradiction with the minimality of the chosen map.

How do you show that you can always find a suitable region to remove? By showing that the minimal map (with its nice properties) has to contain a local configuration of a small number of countries from a set, all of whose elements allow such a reduction. Those who worked on the problem referred to such a set as an "unavoidable set" of "reducible configurations." The challenge was to find such a set.

What made this a problem for the computer age is that the set Appel and Haken found contained almost 1500 configurations. The two researchers spent three years, working with (what was then regarded as) a powerful computer at the University of Illinois, developing a procedure that would generate an unavoidable set of reducible configurations, and writing a routine for proving reducibility that looked like it would work on the kinds of configuration they would encounter. This was very definitely experimental mathematics. Over the course of the three years, the computer outputs led them to make some 500 alterations to their generating procedure. Appel and Haken themselves had to analyze some 10,000 local map configurations *by hand calculation,* and the computer examined a further 2,000 configurations and proved the reducibility of a final total of 1482 configurations in the unavoidable set. When the task was completed in June 1976, it had taken four years of intense work and 1200 hours of computer time. For a more detailed account, we refer the reader to Chapter 7 of Keith Devlin's book *Mathematics: The New Golden Age.*

In the Footsteps of Ramanujan

The great Indian mathematician Srinivasa Ramanujan (1887–1920) whose life has just been substantially fictionalized in *The Indian Clerk* [Leavitt 07] was largely self-taught.[2] Not having been trained in the modern conception of mathematical proof, his approach to mathematical discovery was very much an experimental one, albeit without the aid of a computer (apart from his own very capable brain). He left many remarkable discoveries in his *Notebooks*, about which G. H. Hardy wrote [Hardy 37, p. 147]:

> As Littlewood says "the clear-cut idea of what is meant by a proof, nowadays so familiar as to be taken for granted, he perhaps did not possess at all; if a significant piece of reasoning occurred somewhere, and the total mixture of evidence and intuition gave him certainty, he looked no further."

The nature of Ramanujan's *Notebooks* entries has made the editorial work in explaining his collected results—by Bruce Berndt along with George Andrews and others—a fascinating mixture of experimental and forensic mathematics. "What did he know and how did he know it?" was the central question. The task of understanding just what Ramanujan did is only now drawing to completion, nearly a century after the great Indian's death. We illustrate this state of affairs with a beautiful continued fraction Ramanujan studied:

$$R_\eta(a,b) = \cfrac{a}{\eta + \cfrac{b^2}{\eta + \cfrac{4a^2}{\eta + \cfrac{9b^2}{\eta + \ddots}}}},$$

where $\eta > 0$ and a, b are real (or complex) numbers.

Ramanujan asserted that for appropriate positive a, b, the symmetrization of the fraction is the fraction of the arithmetic and geometric means

$$\frac{\Re_\eta(a,b) + \Re_\eta(b,a)}{2} = \Re_\eta\left(\frac{a+b}{2}, \sqrt{ab}\right).$$

In fact, this is true for all positive a, b, but proving it in generality requires considering complex variables. The first question is, for which complex

---

[2] A superb biography is [Kanigel 91].

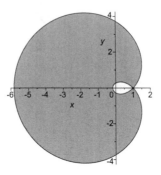

Figure 11. Crandall and Borwein's scatter plot looked just like this.

$c = b/a$ does the continued fraction exist? Crandall and Borwein struggled with this until they decided on the following strategy (for $\eta = 1$): (1) find efficient code to compute $\Re_\eta(a,b)$;[3] (2) pick a criterion to determine whether the fraction is converging; and (3) construct a *scatter plot* of all the c for which this appears to hold [Borwein D. et al. 07]. The virtue of a scatter plot with say 100,000 points is that it does not matter if you misclassify a few hundred. The scatter plot produced the picture shown in Figure 11 so precisely that the equation of the bounding curve could be read off.

The equation of the bounding curve is $|(1+c)/2| \geq \sqrt{|c|}$. For real c, this is the arithmetic-geometric mean inequality, which Figure 11 shows holds for all positive and some negative real numbers since the positive real line is all in the shaded region (though at 1, only just). The precision of the discovery provides substantial reassurance of its truth, but more striking is the fact that the arithmetic and geometric means appear in an answer that came entirely from numerical sampling.

It remained *only* to prove this discovery. Key to this was the need to understand the behavior in the complex plane of the dynamical system $t_0 := 1,\ t_1 := 1$, and

$$t_n := \frac{1}{n} t_{n-1} + \left(1 - \frac{1}{n}\right) \kappa_{n-1} t_{n-2},$$

---

[3]If your code is correct, you should find that

$$\Re_1(1,1) = \log 2, \quad \Re_1(2,2) = \sqrt{2}\left(\tfrac{\pi}{2} - \log(1+\sqrt{2})\right).$$

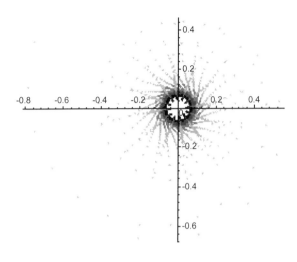

Figure 12. The first 3,000 points of the iteration.

where $\kappa_{2n} := a^2$, $\kappa_{2n+1} := b^2$. This is trivial for $a = b = 1$. Generally, it is surprisingly subtle. If you write code for the iteration numerically, all you see is that the values slosh around to zero.

Surprisingly, convergence is slower the closer a, b are. Indeed convergence is only arithmetic (like $1/n$) when $a = b$.

It turns out that the interesting case reduces to $|a| = |b| = 1$ and that when $a = \pm b$ is purely imaginary, the situation is chaotic. We'll ignore this subcase for now.

A representative picture of the first 3,000 points when $a = e^{\pi i/12}$, $b = e^{\pi i/8}$ is shown in Figure 12. We have ignored the first few points, and the rest progress from light gray to black. Clearly the values spiral in (to zero), as drawing a few more cases shows.

The modulus appeared to be approximately $1/\sqrt{n}$ after n steps, which known theory supported. Plotting $\sqrt{n}t_n$ for $a = e^{\pi i/12}, b = e^{\pi i/8}$ yields the left-hand picture in Figure 13, and plotting $\sqrt{n}t_n$ for $a = e^{\pi i/14}$, $b = e^{\pi i/6}$ gives the right-hand image.

In the first diagram, we see a circle and twelve blobs appearing. In the second, we see 14 blobs and 6 blobs. What is happening?

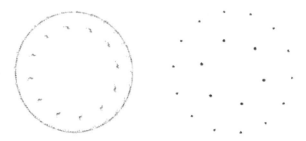

Figure 13. Plots of $\sqrt{n}t_n$.

Figure 14. A plot using different parameters.

Trying a few more cases suggests that the difference is whether a or b is a pth root of unity (in which case it generates p blobs) or not (in which case it converges to a circle). For example, $a = e^{i/12}$, $b = e^{i/8}$ yields the two circular "attractors" in Figure 14.

A year after this investigation was completed, all of these numerical-visual discoveries had been proven by ingenious but traditional human methods, and the behavior of Ramanujan's continued fraction fully explained.

The Robbins Conjecture

In the mid-1930s, Herbert Robbins conjectured that commutativity, associativity, and the single "Robbins axiom"

$$\neg(\neg(x \vee y) \vee \neg(x \vee \neg y)) = x$$

imply all the axioms for a Boolean algebra. It took sixty years to prove this conjecture, and when it happened in 1996, it was done using an automatic theorem prover. This example, while not directly one of experimental mathematics, is another bench mark of the computer's increasing mathematical potency. The same is true for our next and final example of this new kind of mathematics.

The Computation of E_8

The mathematics behind *the exceptional Lie group* E_8 and the details of its 2007 computation are beyond our scope, but its lovely picture is included

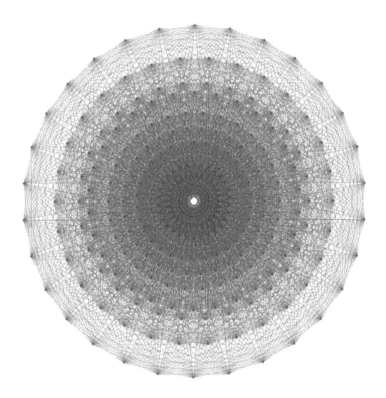

Figure 15. The Gosset polytope associated with E_8.

here. E_8, which is of interest to physicists, has an associated *root system* consisting of 240 vectors in an eight-dimensional space. They form the corners of an eight-dimensional object called the *Gosset polytope*. In the 1960s, Peter McMullen hand-drew a two-dimensional representation. John Stembridge used a computer to replicate the image, obtaining the remarkable picture shown in Figure 15.

The American Institute of Mathematics website[4] says of this (mammoth) computation, "This achievement is significant both as an advance in basic knowledge and because of the many connections between E_8 and other areas, including string theory and geometry."

The computation was both mathematically and computationally sophisticated and generated a huge dataset that, as in any experimental science, interested mathematicians or physicists can now consult and use.

Explorations

1. *Arithmetic progressions of primes.* In 2004, Benjamin Green and 2006 Fields Medalist Terence Tao proved that there are arbitrarily long arithmetic progressions of prime numbers [Green and Tao 08]. In other words they established that for every $n > 0$, there is an integer d for which that every member of the sequence $p, p + d, p + 2d, \ldots, p + nd$ is prime. Their result is a *tour de force* of traditional mathematics, primarily analytic number theory but involving other fields as well. But even here, experimental methods played a role. The authors cite a series of calculations in search of long arithmetic sequences of primes. The longest of these, of length 23, is due to Markus Frind, Paul Underwood, and Paul Jobling, who discovered that the numbers $56211383760397 + 44546738095860k$ are prime for $k = 0, 1, \ldots, 22$. To date, this is the longest known arithmetic progression of primes. Discovery of a longer one is unlikely to secure you a professorship at Harvard (or anywhere else for that matter), but it would almost certainly get you a mention in *Science News*.

2. *Discrete dynamical systems.* These offer a treasure trove of experimental opportunities: symbolic, numerical, and graphical computation can

---

[4]http://www.aimath.org/E8/

each provide insight. You get a "discrete dynamical system" by taking a continuous self-map $f : S \to S$ of a set to itself and iterating:

$$x_0 \in S, \; x_{n+1} := f(x_n) \quad (S).$$

The task is to study the behavior of this system.

When S is a real interval, a famous 1964 theorem by Oleksandr Sharkovsky implies that "period three implies chaos." In the words of Li and Yorke: "If [S] has a point of period three then it has points of all periods" (and much more also follows) [Li and Yorke 75]. How wrong this goes in more general settings is illuminated by the following three very simple systems in the Euclidean plane. In each case you are challenged to determine the behavior of the system.

(a) $a_0 := (x, 0)$, $a_1 := (0, y)$, $a_{n+1} := |a_n| - a_{n-1}$ $(|(u, v)| := (|u|, |v|))$.

(b) $a_0 := x$, $a_1 := y$, $a_{n+1} := \frac{y^2 + a_n^2}{2}$.

(c) $(u, v) \leftarrow (v, u^2 - v^2)$. In other words $u_{n+1} = v_n$ and $v_{n+1} := u_n^2 - v_n^2$.

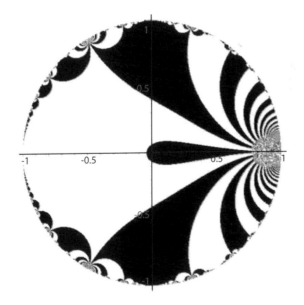

Figure 16. Solution to a complex inequality.

3. *Visualizing an inequality.* Figure 16 shows in black the points of the unit disc in the complex plane where the inequality

$$\left| \frac{\sum_{n=-\infty}^{\infty} q^{(n+1/2)^2}}{\sum_{n=-\infty}^{\infty} q^{n^2}} \right| \leq 1$$

holds. Notice the remarkable level of replication and self-similarity but that every point on the interval [0,1] is black; there is nothing complicated happening on the real line. This is closely related to the Ramanujan arithmetic-geometric mean fraction discussed in this chapter.

Answers and Reflections

Chapter 1

1. *Recognizing sequences.* The sequences can all be found in Sloane's online encyclopedia, along with much more information.

   (a) These are the first few *perfect numbers* (numbers that are equal to the sum of their proper divisors): $6 = 1 + 2 + 3$, $28 = 1 + 2 + 4 + 7 + 14$, etc.

   (b) The *Motzkin numbers.* Among other interpretations, these numbers count the number of ways to join n points on a circle by nonintersecting chords and the number of length n paths (zigzags) from $(0,0)$ to $(n,0)$ that do not go below the horizontal axis and are made up of steps $(1,1)$, $(1,-1)$, and $(1,0)$. The ordinary generating function is $(1 - \sqrt{1 - 2x - 3x^2})/(2x^2)$.

   (c) The *Bell numbers*, whose exponential generating function is $e^{e^x - 1}$.

   (d) Values of *Bell polynomials*, in this case counting ways of placing n labeled balls into n unlabeled (but two-colored) boxes.

   (e) This is *Aronson's sequence*, whose definition is: "t is the first, fourth, eleventh, ... letter of this sentence."

   (f) The number of possible chess games after n moves.

2. *The 3n + 1 problem.* There is a huge literature on the $3n + 1$ problem and much of it can be accessed through the online resources MathWorld, Planet Math, or Jeff Lagarias's survey at http://www.cecm.sfu.ca/organics/papers/lagarias/index.html.

3. *Continued fractions.*
$$\pi = [3, 7, 15, 1, 292, 1, 1, 1, 2, 1, \ldots]$$
$$e = [2, 1, 2, 1, 1, 4, 1, 1, 6, 1, \ldots]$$

Chapter 2

1. *BBP formulas.* A detailed discussion of BBP formulas for logarithms (and for arctangent values) is given in Section 3.6 of *Mathematics by Experiment*. As explained there, it appears there is no such formula for the natural log of 23. Formulas are as follows (b) for π^2 in base 2 and base 3 and (c) for Catalan's constant in binary:

$$\pi^2 = \frac{9}{8} \sum_{k=0}^{\infty} \frac{1}{64^k} \left(\frac{16}{6k+1} + \frac{8}{6k+2} - \frac{2}{6k+4} - \frac{1}{6k+5} \right),$$

$$\pi^2 = \frac{2}{27} \sum_{k=0}^{\infty} \frac{1}{729^k} \left(\frac{243}{(12k+1)^2} - \frac{405}{(12k+2)^2} - \frac{81}{(12k+4)^2} \right.$$

$$- \frac{27}{(12k+5)^2} - \frac{72}{(12k+6)^2} - \frac{9}{(12k+7)^2} - \frac{9}{(12k+8)^2}$$

$$\left. - \frac{5}{(12k+10)^2} + \frac{1}{(12k+11)^2} \right),$$

$$G = \frac{1}{1024} \sum_{k=0}^{\infty} \frac{1}{4096^k} \left(\frac{3072}{(24k+1)^2} - \frac{3072}{(24k+2)^2} - \frac{23040}{(24k+3)^2} \right.$$

$$+ \frac{12288}{(24k+4)^2} - \frac{768}{(24k+5)^2} + \frac{9216}{(24k+6)^2} + \frac{10368}{(24k+8)^2}$$

$$+ \frac{2496}{(24k+9)^2} - \frac{192}{(24k+10)^2} + \frac{768}{(24k+12)^2} - \frac{48}{(24k+13)^2}$$

$$+ \frac{360}{(24k+15)^2} + \frac{648}{(24k+16)^2} + \frac{12}{(24k+17)^2} + \frac{168}{(24k+18)^2}$$

$$\left. + \frac{48}{(24k+20)^2} - \frac{39}{(24k+21)^2} \right).$$

A precise version of our remark about formulas for π is given as Theorem 3.6 in *Mathematics by Experiment*.

2. *There is no known BBP formula for e in any base.* There are strikingly few series for e in the literature as compared to those for π. One reason may

be that the Taylor series for e is so effective that there has been little reason to look any further. For example, to compute e^{100} to high precision, you can instead calculate $e^{100/128}$—which is much more rapidly computable—and then square the answer seven times.

3. *Spigot algorithms for π and e.* The whole issue of normality and digit algorithms is discussed at some length in Chapter 4 of *Mathematics by Experiment*. It is worth emphasizing how little can be proved. It has been conjectured that the binary digits of $\sqrt{2}$ contain asymptotically the same number of 0s and of 1s; but the best proved result is that $O(\sqrt{n})$ of the first n bits must be 1. Likewise, no one can prove that π has infinitely many 7s (or 2s, or ...) in its decimal expansion. But it does!

Chapter 3

1. *Name that number.*

(a) $\sqrt{2} + \sqrt{3}$.

(b) $\sqrt[3]{2} + \sqrt{3}$.

(c) $1 + e^{\pi}$.

(d) $\pi^e - 2$.

(e) $\pi^2 + e\pi - 10$.

(f) The real root of $z^3 - z - 1$. (An alternative answer to (f) is that it is the smallest *Pisot number*, that is,

$$\lim_{n \to \infty} \{n\alpha\} = 0,$$

where, as before, $\{x\}$ denotes the fractional part of x. The golden mean $G := (\sqrt{5} + 1)/2$ also has this property since, with $g := (1 - \sqrt{5})/2$, you get $g^n + G^n = L_n$, where the L_n are integers called the *Lucas numbers* and satisfy the same recursion as the Fibonacci numbers.)

(g) This is the larger of the two real roots of *Lehmer's polynomial*

$$z^{10} + z^9 - z^7 - z^6 - z^5 - z^4 - z^3 + z + 1.$$

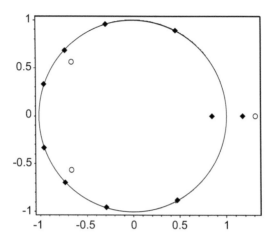

Figure 17. The zeroes of Lehmer's polynomial and of the smallest Pisot number.

In 1933, Lehmer conjectured that it is the smallest *Salem number* (all other roots lie inside or on the unit disc).

Figure 17 shows the zeroes of Lehmer's polynomial (ten diamonds) and of the smallest Pisot number (three circles).

(h) The number has continued fraction $[0, 1, 2, 3, 4, 5, \ldots]$ and turns out to be the ratio of Bessel functions $I_1(2)/I_0(2)$. If you have found the continued fraction, typing "arithmetic continued fraction" into a search engine will probably tell you the rest. Alternatively, entering the digits into Sloane will get you there.

2. *Name that sum.* The sum is $32/\pi^3$. This was discovered by Gourevitch using integer relation methods. No proof is known.

3. *Reflection.* There is a beautiful theorem due to Gelfond and Schneider that, whenever α and β are complex algebraic numbers, with α not 0 or 1 and β irrational, then α^β is transcendental. It follows that e^π is transcendental (because $e^{-\pi/2} = i^i$). By contrast, π^e has never been proved irrational (nor indeed have πe or $e + \pi$).

Chapter 4

1. *Closed forms for $\zeta(2n), \beta(2n+1)$.*

   (a) There are also familiar Fourier series techniques for evaluating the even zeta-values. All techniques break down in the case of odd natural numbers. The closed form, as we saw, is

   $$\zeta(2m) = \frac{(-1)^m (2\pi)^{2m} B_{2m}}{(2(2m)!)},$$

   where B_{2m} is the $(2m)$-th *Bernoulli number*. The generating function of the Bernoulli numbers[5] is $z/(e^z - 1)$. As always, there is plenty of information available in the usual places.

   (b) In this case we have

   $$\beta(2m+1) = \frac{(\pi/2)^{2m+1} |E_{2m}|}{(2(2m)!)},$$

   where the (even) Euler numbers are defined by

   $$\sec(x) = \sum_{n=0}^{\infty} \frac{E_{2n} x^{2n}}{(2n)!},$$

   and start $1, -1, 5, -65, 1365$, as discussed in Chapter 3. Ramanujan found the hyperbolic series identity

   $$\zeta(3) = \frac{7\pi^3}{180} - 2\sum_{k=1}^{\infty} \frac{1}{k^3(e^{2\pi k} - 1)},$$

   in which the hyperbolic series "error" is around -0.003742745, and which, to our knowledge, is the closest you can get to expressing $\zeta(3)$ as a rational multiple of π^3.

2. *Multi-dimensional zeta functions.* $\zeta(2,1) = \zeta(3)$. Borwein and David Bradley found 32 proofs of this [Borwein and Bradley 06]. They illustrate diverse combinatorial, algebraic and analytic approaches to multi-

---

[5]The only nonzero odd Bernoulli number is $B_1 = -1/2$. The even numbers start $1, 1/6, -1/30, 1/42, -1/30, 5/66, \ldots$

ple zeta-values. Perhaps the easiest proof is by telescoping, where you write

$$S := \sum_{n,k>0} \frac{1}{nk(n+k)} = \sum_{n,k>0} \frac{1}{n^2}\left(\frac{1}{k} - \frac{1}{n+k}\right)$$

$$= \sum_{n=1}^{\infty} \frac{1}{n^2} \sum_{k=1}^{n} \frac{1}{k} = \zeta(3) + \zeta(2,1).$$

On the other hand, by symmetry, we also have

$$S = \sum_{n,k>0} \left(\frac{1}{n} + \frac{1}{k}\right) \frac{1}{(n+k)^2}$$

$$= \sum_{n,k>0} \frac{1}{n(n+k)^2} + \sum_{n,k>0} \frac{1}{k(n+k)^2} = 2\zeta(2,1),$$

and we are done.

3. *The Riemann hypothesis.*

   (a) There are six zeroes on the requested interval, as shown in Figure 18.

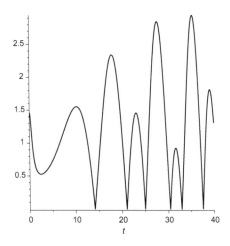

Figure 18. The six zeroes on the given interval.

To 20 decimal places, the zeroes are

14.134725141734693790, 21.022039638771554993,
25.010857580145688763, 30.424876125859513210,
32.935061587739189691, 37.586178158825671257.

You should plot the function first, since the zero-finder in a computer package will almost certainly need to be given some help in localizing the zeroes and in confirming you have not *missed* any.

The first 1.5 billion zeroes are known to lie on the critical line,[6] as indeed are all whose imaginary part is less than 10^{13}. Sadly, the "Law of small numbers" still rules at that size; Andrew Odlyzko, who has computed twenty billion zeroes around 10^{23}, has suggested that you would have to have vastly more numerical confirmation to be firmly convinced of the truth of RH.[7] What does "vastly" mean here? According to Odlyzko, you should take his figure of 10^{23} and exponentiate twice.

(b) You should get a picture somewhat like the one in Figure 19.

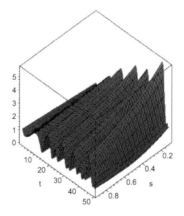

Figure 19. Do you get a picture like this?

---

[6]Gourdan's 2004 paper http://numbers.computation.free.fr/Constants/Miscellaneous/zetazeros1e13-1e24.pdf records current, highly refined computational methods for finding the zeroes.
[7]See http://www.dtc.umn.edu/~odlyzko/zeta_tables/.

This has the surprising feature that all the cross sections appear to be of three types: monotone decreasing, monotone increasing, or decreasing and then increasing (which covers the other two cases). This was discovered in 2002 in an undergraduate complex variable course.[8] If you could prove it, you would gain instant fame in the mathematical community, your name would go down in history, and you would win a million dollars. It is equivalent to the Riemann hypothesis!

Chapter 5

1. *Integration.* The seven answers are

   (a) $4\pi \log^2 2 + \pi^3/3$

   (b) $\pi(\log 2)/8 + G/2$,

   (c) $(\log 2)/2 + \pi^2/32 - \pi/4 + \pi(\log 2)/8$,

   (d) $\pi \log(2)$,

   (e) $\log \pi - 2 \log 2 - \gamma$,

   (f) $\pi\sqrt{2}$,

   (g) $\pi^3/12 + 2\pi \log 2 + \pi^3/3 \log 2 - 3\pi/2\zeta(3)$.

Euler's constant is defined by

$$\lim_{n \to \infty} \left(1 + \frac{1}{2} + \ldots + \frac{1}{n}\right) - \log n,$$

which means showing the limit exists. It has not been proved to be irrational, but if it is, the numerator has at least ten million digits. This would be fantastic, as the 20-odd characters in the definition would then encode two very large integers. A nice integral representation is

$$\gamma = \int_0^\infty \left(\frac{1}{e^t - 1} - \frac{1}{te^t}\right) dt.$$

---

[8]See [Saidak and Zvengrowski 03].

Chapter 6

Did you make any progress with that inequality

$$2 + \frac{2}{45}x^3 \tan x > \frac{\sin^2 x}{x^2} + \frac{\tan x}{x} > 2 + \frac{16}{\pi^4}x^3 \tan x > 2$$

where $0 < x < \pi/2$?

This example turned out to be more illustrative than we originally intended, as we now describe. *Wilker's inequality*, taken from *Experimentation in Mathematics*, cannot be correct as typed, since $2/45 < 16/\pi^4$.

Hunting on the Web[9] revealed that it should be

$$2 + \frac{8}{45}x^3 \tan x > \frac{\sin^2 x}{x^2} + \frac{\tan x}{x} > 2 + \frac{16}{\pi^4}x^3 \tan x > 2,$$

and we see that the "best possible" constant $8/45$ yields the correct series at zero since

$$\frac{\sin^2 x}{x^2} + \frac{\tan x}{x} = 2 + \frac{8}{45}x^4 + O(x^6).$$

We also see the inequalities are rather good if we stay away from $\pi/2$.

Computing the Taylor series is another way to "error correct." In what sense is $16/\pi^4$ best possible? (Note that $8\pi^4/16.45 = \zeta(4)$.) See Figure 20.

How did you get on with the sum

$$\sum_{n=0}^{\infty} \frac{(4n+3)}{(n+1)^2} \frac{\binom{2n}{n}^2}{2^{4n}}?$$

The answer is 1. The finite sum of the first N terms is

$$1 - \binom{2N}{N}^2 / 2^{4N},$$

which can be proved by induction.

---

[9]It is still much easier to hunt for named objects such as "Wilker's inequality" than mathematical expressions, and it is likely to remain so for quite a while.

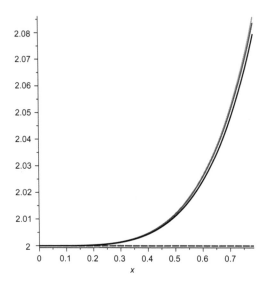

Figure 20. The four functions ($x = 2$ included) in Wilker's inequality.

Chapter 7

1. *Fast arithmetic.*

   (a) The iteration for reciprocation is $x_{n+1} = x_n(2 - x_n b)$.

   (b) The iteration $x_{n+1} = x_n(3 - bx_n^2)/2$ produces $1/\sqrt{b}$ without use of division. Now multiply by b to obtain $\sqrt{b}$. Thus, it is realistic to view reciprocation as three or four times as hard as multiplication and extracting a square root as six times as hard.

2. *Monte Carlo calculation of* π. In past times, the pseudorandom number generators on desktop computers were often very poor, so while a Monte Carlo computation was a poor way to discover digits of π, it was a fine way to uncover problems with a built-in random number generator.

3. *Convergence rates.* The iterations converge to e^{π}. The iteration in (a) does so quadratically (you must work to roughly twice the intended precision), the one in (b) woefully slowly. The former illustrates that

e^{π} is the easiest transcendental constant to fast-compute. You can re-express the limit as a fast product

$$e^{\pi} = 32 \prod_{k=0}^{\infty} \left(\frac{1 + x_k}{2} \right)^{2^{-k}},$$

where

$$x_{k+1} = \frac{2\sqrt{x_k}}{1 + x_k} \quad \text{and} \quad x_0 := \frac{1}{\sqrt{2}}.$$

The fast algorithms for π cleverly manage to "take a logarithm."

Chapter 8

1. *The Psi function.* In the same fashion,

$$\zeta(\overline{2}, 1) := \sum_{n=1}^{\infty} (-1)^n \frac{H_{n-1}}{n^2} = \frac{\zeta(3)}{8}.$$

A rabbit-out-of-the-hat proof is provided by defining

$$J(x) := \sum_{n=1}^{\infty} H_{n-1} x^n / n^2$$

and asserting that J satisfies the functional equation

$$J(-x) = -J(x) + \frac{1}{4}J(x^2) + J\left(\frac{2x}{x+1}\right) - \frac{1}{8}J\left(\frac{4x}{(x+1)^2}\right).$$

Now set $x = 1$ and obtain $\zeta(\overline{2}, 1) = J(-1) = J(1)/8 = \zeta(2,1)/8 = \zeta(3)/8$.

To prove the functional equation, just differentiate and simplify. As for *discovering* it, that requires deciding to look for a relation in the right form. Then you can use integer relation methods to find it.

2. *What is κ?* The floating-point value suggests nothing, but there *might* be a pattern in the continued fraction. Indeed, the Gauss-Kuzmin distribution for a random real number expects about 41% of the entries to be 1. Redoing the computations to 50 places yields

$$[2, 3, 10, 7, 18, 11, 26, 15, 34, 19, 42, 23, 50, 27, 58, 31, 66, 35, 73].$$

Ignoring the last entry, as you almost always should, a pattern jumps out. You see two interspersed arithmetic sequences. In particular, this means the continued fraction is not periodic, so κ is not a quadratic irrationality[10] and in consequence (once the fraction is proved) shows $e^{\sqrt{2}}$ is irrational.

Chapter 9

1. *Finding limits.*

   (a) The limit when it exists, independent of a, is $(\sqrt{5}-1)/2$. The limit exists in an interval.

   (b) The limit is $(\sqrt{13}-1)/2$.

   (c) The limit is $\pm\tan(\theta)$.

2. *Hirschhorn's limit.* The limit is $2/3$.

3. *Mean iterations.*

   (a) The limit is $L(a,b) = \sqrt{ab}$.

   (b) The limit is $L(a,b) = a - b/\log(a/b)$ when a and b differ (this is correctly interpreted as a limit when $a = b$). In each case, a proof consists of arguing as follows.

   The limit is always mean and so is the unique mean satisfying the *invariance principle* $L(a_n, b_n) = L(a_{n+1}, b_{n+1})$ since then

   $$L(a,b) = L(a_n, b_n) = L(a_{n+1}, b_{n+1})$$
   $$= L(\lim_{n\to\infty} a_n, \lim_{n\to\infty} b_n) = (M \otimes N)(a,b),$$

   where the last equality follows because L is a mean. The invariance is fun to check in a computer algebra system.

---

[10] It is a famous result of Lagrange that a nonperiodic nonterminating continued fraction cannot arise from a quadratic or rational number. Try typing "periodic continued fraction" into a search engine.

The same approach will allow you to show that Archimedes' iteration discussed in Chapter 7 is a special case of an asymmetric mean iteration. Concisely, you should be able to show that with

$$H(a,b) := (2ab)/(a+b), N(a,b) := \sqrt{b(a+b)/2}$$

you get

$$H \otimes N(a,b) = \frac{\sqrt{b^2 - a^2}}{\arccos(a/b)},$$

for $0 < a < b$, by using the invariance principle. By showing that

$$I(a,b) := \int_0^{\pi/2} \frac{dt}{\sqrt{a^2 \cos^2(t) + b^2 \sin^2(t)}}$$

satisfies $I(a,b) = I((a+b)/2), \sqrt{ab})$, the invariance principle also establishes that $A \otimes G(a,b) = \frac{\pi}{2}/I(a,b),$[11] where, as usual, A and G are the arithmetic and geometric mean functions.

Chapter 10

1. *Simplification.* The real numbers

$$\alpha_1 = \sqrt[3]{\frac{3^{5/3} - 3}{2}} \quad \text{and} \quad \alpha_2 = -\sqrt[3]{\frac{3.7^{1/3} - 5}{2}}$$

can both be found by using the ISC or a minimal polynomial calculation. In the second case, you may be better off hunting for the cube. If you typed the expression into a CAS, it may well have returned the complex cube root of a negative number! You could have removed the ambiguity by writing

$$\alpha_1 := \sqrt[3]{\cos(2\pi/9)} + \sqrt[3]{\cos(4\pi/9)} - \sqrt[3]{\cos(\pi/9)}$$

in the first case, but human mathematics is full of ambiguity.

---

[11]It is nice challenge to establish this invariance in a CAS.

2. *Recent American Mathematical Monthly problem.* Folkmar Bornemann's nice solution in Maple 9.5 shows first that $\sigma(m,n)$ is independent of n, and second that $\sigma(m,m) = 4^m$. Unfortunately, in Maple 10 and Maple 11, a bug leads to $\sigma(m,m) = 3.4^m$ (Yes, that would mean $4 = 12$!). Mathematica needs much more coaxing but avoids making the same error.

Chapter 11

2. *Discrete dynamical systems.*

(a) For every point (x, y) we cycle in nine steps. This can be discovered by numerically iterating, but a better approach is to plot the orbits, as in Figure 21.

It can actually be *proved* by composing the map symbolically with itself nine times and simplifying the result carefully in a CAS. For

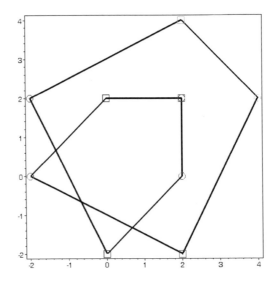

Figure 21. The four symbols show the effect of plotting 20 points from four initial values. We only see nine of each. A periodic orbit can be seen starting at $(2, 2)$, with alternating circles and squares.

example, in Maple

```
d1:=proc(x,y,N) local n, u; u[0]:=[x,0]; u[1]:=[0,y];
for n to N-1 do
    u[n+1]:=[abs(u[n][1])-u[n-1][1],
             abs(u[n][2])-u[n-1][2]]
od; end;
```

will output the Nth iterate. Then, we run

```
> d1(x,y,9);
```

$$[|-|-||||x|+x|-|x||-|x|-x|+||x|+x|-|x|||||x|+x|-|x||$$
$$-|x|-x|,|-|-|||-|-|y|+y|+|y||-|y|+y|-|-|y|+y|+$$
$$|y||+|-|-|y|+y|+|y||-|y|+y|+||-|-|y|+y|+|y||-|y|$$
$$+y|-|-|y|+y|+|y||-|||-|-|y|+y|+|y||-|y|+y|-|-|$$
$$y|+y|+|y||+|-|-|y|+y|+|y||-|y|+y]$$

```
> simplify(d1(x,y,9)) assuming x>0,y>0;
```

$$[x,0]$$

```
> simplify(d1(x,y,10)); assuming x>0,y>0;
```

$$[0,y]$$

We may likewise check that the three other sign choices perform similarly.

Alternatively, using the matrices to represent the iteration (how?)

$$A := \begin{pmatrix} 0 & 1 \\ -1 & 1 \end{pmatrix}, \quad B := \begin{pmatrix} 0 & 1 \\ -1 & -1 \end{pmatrix},$$

you discover that $B^3 = I = -A^3$ and that the *symbolic dynamics* of the system reduce to showing that any possible word of "moves" reduces to a string of the form $A^3 B A^3 B^2 = I$.

(b) Each iterate is a polynomial in x and y. The sixth fills a screen or two when expanded. If the limit λ exists, it must satisfy

$$1 - y^2 = (1 - \lambda)^2.$$

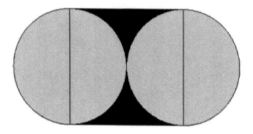

Figure 22. Convex hull of the circles with centers at $(\pm 1, 0)$.

The iteration converges if and only if $(x, y) \in C$, where C pictured in Figure 22 is the convex hull of the circles with centers at $(\pm 1, 0)$.

To discover this result, try writing a couple of lines of code that (i) stores the points $(i/N, j/N)$ for, say, $0 \le i, j < 4N$, and for which the first M iterates stay in a prescribed interval, or even the unit interval, and (ii) plots these points. Depending on the grid size, it should look somewhat like the pictures in Figure 23. (We appealed to symmetry and only examined positive pairs.) The overflow in the first picture is reflective of the slower convergence around $(0, 1)$.

(c) Here are three assertions to prove. (i) There is no point of period two. (ii) There are points of period three (so "period three implies chaos" is not entirely true in the plane). (iii) If the initial values lie in the open unit square (with $|x| < 1$, $|y| < 1$), then the iteration converges to the origin.

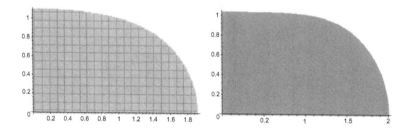

Figure 23. Did your output look like this?

Figure 24. Life can get complicated.

Elsewhere, life gets complicated. It can be proved that there are divergent orbits. The composite picture in Figure 24 shows what *appears* to happen if we plot the points that have converged—necessarily to zero—after many iterations, with darker colors indicating that the convergence was slower and black indicating nonconvergence. This example is due to Marc Chamberland.

Final Thought

It is perfectly reasonable to suppose that many true statements of mathematics have no proof or nothing that we could ever convince ourselves was a proof. More broadly, the experience gained from working experimentally has shown little, if any, correlation between the difficulty of *discovering* something and the difficulty of subsequently *proving* it. Indeed, there is no *a priori* reason why there should be any such correlation.

Along this line, Gregory Chaitin has conjectured a "heuristic principle" that the theorems of a finitely specified theory cannot be significantly more complex than the theory itself. Recently, Cristian Calude and Helmut Jürgensen proved a fairly strong form of this conjecture. They showed that the theorems of a consistent and finitely specified theory that is arithmetically sound (e.g., Peano Arithmetic or Zermelo-Frankel set theory) have bounded complexity [Calude and Jürgensen 05]. As a result, *the probability that a true sentence is provable in the theory tends to zero when its length tends to infinity, while the probability that a sentence of any fixed length is true is strictly bounded above zero.*

Thus, we are left to speculate on how the empirical methods that we have discussed can best be seen as tools to explore this realm of true but unprovable results, or even of true but very-difficult-to-prove results. Here are a few, well-known, number-theoretic candidates:

1. Are there any odd perfect numbers? It is known that any odd perfect number must have a prime divisor larger than one hundred million. (It took ten CPUs four months to determine this.)

2. Are there are infinitely many even perfect numbers, or equivalently infinitely many Mersenne primes? See http://research.att.com/~njas/sequences/A000043.

3. Is Lehmer's conjecture true? It states that $\varphi(n)|n-1$ if and only if n is prime, where $\varphi(n)$ is Euler's *totient function*, which counts the

Aristotle and Plato
just walking away
into the setting sun

number of numbers less than n that are relatively prime to n. It is known that any counterexample must have at least 15 odd prime factors and is very large. (If $3|n$, there are at least a quarter of a million prime factors.)

Such questions need some careful thought in a field that, due to the hitherto unstoppable march of Moore's Law, seems destined to become ever more dependent on computational exploration.

In June 2008 a National Nuclear Security Administration supercomputer, the Roadrunner, built by IBM, achieved an operational rate of a petaflop—10^{15} arithmetic operations per second. This is more than 200 billion times the speed of ENIAC only 60 years ago. It happened two years earlier than suggested by quite recent predictions. Interestingly, Roadrunner uses processors similar to those used in game stations. Mathematics, *mutatis mutandis*, will sooner or later experience profound change.

Additional Reading and References

The following list includes not only books and papers explicitly mentioned in the book, but a number of (largely recent) books at various levels that you may find useful.

[Adams 94] Colin Adams. *The Knot Book*. New York: W. H. Freeman, 1994.

[Bailey et al. 97] D. H. Bailey, P. B. Borwein, and S. Plouffe. "On the Rapid Computation of Various Polylogarithmic Constants." *Mathematics of Computation* 66 (1997), 903–913.

[Bailey et al. 07] David Bailey, Jonathan Borwein, Neil J. Calkin, Roland Girgensohn, D. Russell Luke, and Victor H. Moll. *Experimental Mathematics in Action*. Wellesley, MA: A K Peters, 2007.

[Bornemann et al. 04] Folkmar Bornemann, Dirk Laurie, Stan Wagon, and Jörg Waldvogel. *The SIAM 100-Digit Challenge: A Study in High-Accuracy Numerical Computing*. Philadelphia: Society for Industrial Mathematics, 2004.

[Boros and Moll 04] George Boros and Victor Moll. *Irresistible Integrals: Symbolics, Analysis, and Experiments in the Evaluation of Integrals*. New York: Cambridge University Press, 2004.

[Borwein and Bailey 08] Jonathan Borwein and David Bailey. *Mathematics by Experiment: Plausible Reasoning in the 21st Century*, second edition. Wellesley, MA: A K Peters, 2008.

[Borwein and Borwein 87] Jonathan M. Borwein and Peter B. Borwein. *Pi and the AGM: A Study in Analytic Number Theory and Computational Complexity*. New York: John Wiley & Sons, 1987 (Paperback, 1998).

[Borwein and Bradley 06] J. M. Borwein and D. M. Bradley. "Thirty Two Goldbach Variations." *International Journal of Number Theory* 21 (2006), 65–103.

[Borwein et al. 04] Jonathan Borwein, David Bailey, and Roland Girgensohn. *Experimentation in Mathematics: Computational Paths to Discovery*. Natick, MA: A K Peters, 2004.

[Borwein D. et al. 07] D. Borwein, J. Borwein, R. Crandall, and R. Mayer. "On the Dynamics of Certain Recurrence Relations." *Ramanujan Journal* (Special issue for Richard Askey's 70th birthday) 13:1–3 (2007), 63–101.

[Borwein P. et al. 07] Peter Borwein, Stephen Choi, Brendan Rooney, and Andrea Weirathmueller. *The Riemann Hypothesis: A Resource for the Afficionado and Virtuoso Alike.* New York: CMS–Springer Books, 2007.

[Calude 07] Christian S. Calude. *Randomness and Complexity: From Leibniz to Chaitin.* Singapore: World Scientific Press, 2007.

[Calude and Jürgensen 05] Cristian S. Calude and Helmut Jürgensen. "Is Complexity a Source of Incompleteness?" *Advances in Applied Mathematics* 35 (2005), 1–15.

[Chaitin and Davies 07] Gregory Chaitin and Paul Davies. *Thinking About Gödel and Turing: Essays on Complexity, 1970–2007.* Singapore: World Scientific, 2007.

[Crandall and Pomerance 05] Richard Crandall and Carl Pomerance. *Prime Numbers: A Computational Perspective*, second edition. New York: Springer, 2005.

[Davis 06] Philip J. Davis. *Mathematics and Common Sense: A Case of Creative Tension.* Wellesley, MA: A K Peters, 2006.

[Devlin 97] Keith Devlin. *Mathematics: The Science of Patterns.* New York: Henry Holt, 1997.

[Devlin 99] Keith Devlin. *Mathematics: The New Golden Age.* New York: Columbia University Press, 1999.

[Finch 03] Stephen R. Finch. *Mathematical Constants.* Cambridge, UK: Cambridge University Press, 2003.

[Franco and Pomerance 95] Z. Franco and C. Pomerance. "On a Conjecture of Crandall Concerning the $3n + 1$ problem." *Mathematics of Computation* 64 (1995), 1333–1336.

[Giaguinto 07] Marius Giaguinto. *Visual Thinking in Mathematics.* Oxford, UK: Oxford University Press, 2007.

[Gold and Simons 08] Bonnie Gold and Roger Simons, editors. *Proof and Other Dilemmas: Mathematics and Philosophy.* Washington, DC: Mathematical Association of America, 2008.

[Goldstein 73] L. J. Goldstein. "A History of the Prime Number Theorem." *American Mathematical Monthly* 80 (1973), 599–615.

[Graham et al. 94] Ronald L. Graham, Donald E. Knuth, and Oren Patashnik. *Concrete Mathematics.* Boston: Addison-Wesley, 1994.

[Green and Tao 08] Ben Green and Terence Tao. "The Primes Contain Arbitrarily Long Arithmetic Progressions." *Annals of Mathematics* 2 167:2 (2008), 481–547.

[Guy 94] Richard K. Guy. *Unsolved Problems in Number Theory.* Heidelberg: Springer-Verlag, 1994.

[Hales 05] T. C. Hales. "A Proof of the Kepler Conjecture." *Annals of Mathematics* 162 (2005), 1065–1185.

[Hardy 37] G. H. Hardy. "The Indian Mathematician Ramanujan." *American Mathematical Monthly* 44 (1937), 137–155.

[Havil 03] J. Havil. *Gamma: Exploring Euler's Constant.* Princeton, NJ: Princeton University Press, 2003.

[Hersh 99] Reuben Hersh. *What is Mathematics Really?* Oxford, UK: Oxford University Press, 1999.

[Kanigel 91] Robert Kanigel. *The Man Who Knew Infinity: A Life of the Indian Genius Ramanujan.* New York: Scribner, 1991.

[Koecher 80] Max Koecher. "Letter" (in German). *Math Intelligencer* 2:2 (1980), 62–64.

[Krantz 09] Steven G. Krantz. *The Proof Is in the Pudding: A Look at the Changing Nature of Mathematical Proof.* To appear from Springer-Verlag, 2009.

[Lakatos et al. 76] Imre Lakatos, John Worrall, and Elie Zahar, editors. *Proofs and Refutations: The Logic of Mathematical Discovery.* Cambridge, UK: Cambridge University Press, 1976.

[Leavitt 07] David Leavitt. *The Indian Clerk.* New York: Bloomsbury, 2007.

[Li and Yorke 75] Tien-Yien Li and James A. Yorke. "Period Three Implies Chaos." *American Mathematical Monthly* 82:10 (1975), 985–992.

[MacHale 93] Desmond MacHale. *Comic Sections: Book of Mathematical Jokes, Humour, Wit and Wisdom.* Dublin: Boole Press, 1993.

[Perko 74] Kenneth A. Perko. "On the Classifications of Knots." *Proceedings of the American Mathematical Society* 45 (1974), 262–266.

[Petkovsek et al. 96] Marko Petkovsek, Herbert Wilf, and Doron Zeilberger. $A = B$. Natick, MA: A K Peters, 1996.

[Riemann 59] Bernhard Riemann. "Über die Anzahl der Primzahlen unter einer gegebenen Grösse" ("On the number of primes less than a given quantity"). *Monatsberichte der Berliner Akademie*, November, 1859.

[Rolfsen 76] Dale Rolfsen. *Knots and Links.* Houston: Publish or Perish, Inc., 1976.

[Saidak and Zvengrowski 03] Filip Saidak and Peter Zvengrowski. "On the Modulus of the Riemann Zeta Function in the Critical Strip." *Mathematica Slovaca* 53:2 (2003), 145–172.

[Schecter 98] Bruce Schecter. *My Brain Is Open.* New York: Simon and Schuster, 1998.

[Sinclair et al. 07] Nathalie Sinclair, David Pimm, and William Higginson, editors. *Mathematics and the Aesthetic: New Approaches to an Ancient Affinity*, CMS Books in Mathematics. New York: Springer-Verlag, 2007.

[Stanley 99] Richard P. Stanley. *Enumerative Combinatorics*, Vols. 1 and 2. New York: Cambridge University Press, 1999.

[Steele 04] J. M. Steele. *The Cauchy-Schwarz Master Class*. Washington, DC: Mathematical Association of America, 2004.

[Stromberg 81] Karl R. Stromberg. *An Introduction to Classical Real Analysis*. Belmont, CA: Wadsworth, 1981.

[Tao 06] Terence Tao. *Solving Mathematical Problems*. New York: Oxford University Press, 2006.

[Temme 96] Nico M. Temme. *Special Functions: An Introduction to the Classical Functions of Mathematical Physics*. New York: John Wiley, 1996.

[Villegas 07] Fernando R. Villegas. *Experimental Number Theory*. New York: Oxford University Press, 2007.

Index